SPACE EXPLORATION

A Reference Handbook

SPACE EXPLORATION

A Reference Handbook

Mrinal Bali

Introduction by Dr. Harrison H. Schmitt

CONTEMPORARY WORLD ISSUES

ABC-CLIO

Santa Barbara, California
Oxford, England

Library of Congress Cataloging-in-Publication Data

Bali, Mrinal, 1953–
    Space exploration : a reference handbook / Mrinal Bali ; introduction by Harrison H. Schmitt.
        p. cm. — (Contemporary world issues)
    Includes bibliographical references and index.
    1. Astronautics—Handbooks, manuals, etc.   I. Title.   II. Series.
    TL794.B34 1990   919.9'04—dc20   90-19204

ISBN 0-87436-578-3 (alk. paper)

97 96 95 94 93 92 91 90   10 9 8 7 6 5 4 3 2 1

ABC-CLIO, Inc.
130 Cremona Drive, P.O. Box 1911
Santa Barbara, California 93116-1911

Clio Press Ltd.
55 St. Thomas' Street
Oxford, OX1 1JG, England

This book is printed on acid-free paper ∞.
Manufactured in the United States of America

To the next generation:
Sonia, Shivika, Shriya, Sunaina,
Rajni, Gopu, both Bhawnas, Swati,
Saloni, Anjani, Ajay, Sanjay,
Bunty, and Guriya

Contents

Preface

FOR MORE THAN 30 YEARS, HUMAN BEINGS have been launching objects into space to explore its wonders, mysteries, and potential uses. The Soviet satellite *Sputnik 1* was the first such object. Launched on October 4, 1957, it remained in orbit for 96 days and circled Earth 1,400 times. Since then, men have walked on the Moon and *Voyager 1* and *2* have traveled through and beyond the solar system, sending thousands of pictures and voluminous data back to Earth-bound scientists. Despite the immense cost and frustrating, sometimes tragic setbacks, humanity continues to push farther and deeper into the vastness of space.

The distinction between exploration and astronomy is blurred in most books on space, which tend to treat the two subjects as one. There is, however, a difference: astronomy is a visual or mathematical analysis of celestial phenomena, while exploration is the sending of vessels and probes into space. This book is about the physical exploration of space by human beings. It is intended to offer students and researchers access to information on space exploration in a pure and concentrated form, without dilution by related but essentially separate subjects such as astronomy and aeronautics.

Where Does Space Begin?

Space is often thought to be a total vacuum, but Earth's atmosphere extends as far as 800 kilometers (500 miles), at which altitude, though, it becomes so rarefied as to be nonexistent. Most space vehicles orbit well within that radius, and their orbits constantly decay due to friction with the atmosphere. For the purposes of space exploration, *space* is defined as beginning at the altitude at which an orbit can be sustained over a significant length of time—about 320 kilometers (200 miles), which is the average altitude of shuttle missions. This book is a compilation of information on manned and unmanned missions that achieved— or at least attempted to achieve—Earth orbit or went beyond it.

How This Book Is Organized

This book is intended to serve as both a primary information source and a guide to further research. Chapter 1 is an introduction by Dr. Harrison H. Schmitt, who in 1972 went to the Moon on *Apollo 17*. Chapter 2 is a chronology of key events in space exploration. Chapter 3 contains biographical sketches of key individuals, including those who made space exploration possible and the astronauts and cosmonauts who risked their lives to become the first space explorers. Chapter 4 is a tabular listing of all of humanity's ventures into space. Chapter 5 describes the key organizations and industries involved in space exploration. Chapter 6 is an annotated bibliography of printed reference materials, and Chapter 7 lists films and videos on space exploration. A glossary of important terms completes the volume.

Acknowledgments

A book is never written by the author alone. There are always others who contribute to its development by making helpful suggestions, spoken at a gas station or over coffee in a cafeteria or over a long-distance call in the middle of the night by a friend who just got a good idea for the book. Several such friends have contributed to this book. Foremost among them is Brent Hoage, a man with by far the keenest eye for detail that I have ever seen in a budding engineer, and then Tom Stroud, who I hope will work on his own book as keenly as he took interest in this one. I also appreciate the guidance of the editorial staff at ABC-CLIO, particularly that of Laurie Brock. Finally, my deepest regards to the one who typed most of the enormous information in this book and then spent hours alone patiently while I worked on the manuscript: my wife, Neelam.

Introduction
Dr. Harrison H. Schmitt

Dr. Harrison H. Schmitt is a geologist by education and experience. In 1972, he walked on the Moon as an astronaut on the Apollo 17 *lunar mission, and from 1977 to 1983, he was a U.S. Senator from New Mexico. Dr. Schmitt currently consults on space matters and serves as a corporate director for businesses involved in space technology, manufacturing, mining, biotechnology, and banking.*

In the Beginning

I believe that this nation should commit itself to achieving the goal, before this decade is out, of landing a man on the Moon and returning him safely to Earth. No single space project in this period will be more impressive to mankind, or more important in the long-range exploration of space; and none will be so difficult or expensive to accomplish.

President John F. Kennedy,
address to Congress, May 25, 1961

Eight years later, on July 20, 1969, Neil Armstrong, commander of the U.S. spacecraft *Apollo 11,* became the first human to set foot on another body in space—the Moon—and he broadcast the message, "That's one small step for man, one giant leap for mankind."

Humankind's first explorations of the Moon and of space near Earth were also our first steps of evolution into the solar system and eventually into the Milky Way Galaxy, thus confirming what the ancestors of the Pueblo Indians once said, "We walk on Earth, but we live in the sky."

Our ancestors were the earliest explorers of the sky. They took their eyes and minds into space and thus commenced the long and fundamental process of transplanting civilization into space. Over the course of history, this fundamental process has continued as human beings have also gained new insight into themselves and their first planetary home.

The era of space exploration—of humankind's physical excursion into space—began on October 4, 1957, when the U.S.S.R. launched *Sputnik 1*, the first artificial satellite of Earth. *Sputnik 1* shocked the world. Its impact on the world's view of the future was profound. There was wonder at this first reach by humankind into the ocean of space, and there also came some fear. There was fear that the then-oppressive Soviet political system would now dominate the world through its technological prowess. *Sputnik 1* thus led to the correlation of political domination with success in space, and success in space was related to technological capability. Kennedy's challenge to go to the Moon solidified this relationship, as did emphasis in both the United States and the U.S.S.R. on ballistic missiles—weapons that are propelled by rockets into the upper atmosphere—as the cornerstones of national security.

Space exploration thus began in earnest. The late 1950s through the 1960s brought the robotic age of the Pioneer, Ranger, Surveyor, and Mariner space probes from the United States, and the Viking, Luna, Lunakhod, and Kosmos probes from the U.S.S.R. The mid-1960s through the 1970s brought the human age of the U.S. Apollo and Skylab and the Soviet Soyuz and Salyut missions. For many on Earth, these missions fired motivations comparable to those experienced 200 years ago by the free inhabitants of a youthful United States: a frontier beckoned, and the country's leadership facilitated access to that frontier for those with the courage and ambition to build and live there. Both these crossroads in history, the opening of space and the opening of the American frontier, illustrate that the periods during which meaningful steps can be taken toward the creation of new civilizations and human opportunities occur episodically, when scientific, psychological, and political creativity coincide with unwavering leadership.

Under President John F. Kennedy's leadership, the United States' first response to the challenge of space appeared to reflect the arrival at such a crossroads in history. Consequently, at the peak of the Apollo missions, the United States verged on the establishment of bases on the Moon, research stations in Earth orbit, and a new, realistic goal of a foothold on Mars by the end of this century. In fact, the motto of *Apollo 17*, the last U.S. mission to the Moon's surface, was: "The end of the beginning."

The opportunities created by the Apollo program and its generation, however, passed by. Consequently, the responsibility to re-ignite Kennedy's torch for space exploration falls on others.

The Information Age

The benefits which a satellite system should make possible within a few years will stem largely from a vastly increased capacity to exchange information cheaply and reliably with all parts of the world by telephone, telegraph, radio and television. The ultimate result will be to encourage and facilitate world trade, education, entertainment and new kinds of professional, political and personal discourse which are essential to healthy human relationships and international understanding.

President John F. Kennedy,
August 31, 1962,
upon signing into law
the Communications Satellite Act of 1962

The first communications satellites, Comsats, began to orbit Earth in the early 1960s, and since then, evolutions in the collection and distribution of information worldwide have made a distinct impression on the human psyche. The most graphic demonstration of this change came on Christmas Eve, 1968, when hundreds of millions of humans on Earth had new thoughts about a familiar object in the night sky—the Moon. That Christmas Eve, *Apollo 8*, carrying the first three humans ever to leave Earth for another celestial body, arrived at the Moon. The Moon got its first truly live close scrutiny, and through communications satellites, the men of *Apollo 8* shared their view from the spacecraft instantly with their home planet.

The Moon would thus never be the same for anyone, and now we should realize that Earth also will never be the same. Communications and information technologies offer solutions to many of the age-old problems of the human condition.

The broad-scale use of the unique view of Earth offered by satellites began in the 1960s with the development of weather and communications satellites. These pioneering efforts culminated with satellite systems such as Comsat, Intelsat, Inmarsat, et al., and their obvious success rested on the long-established need for and use of global communications and weather information, and the availability of capital incentives and the institutional means to apply such services.

While communications and weather systems proliferated relatively quickly, the world's need for information on Earth's geological and agricultural resources has grown slowly in spite of the obvious value of these resources. Unlike communications and weather data, remotely gathered information on materials and human activities on Earth were not in general use prior to the space age. Such information was generally available as relatively primitive aerial pictures of use only to highly specialized professionals. Now, however, with the existence and ever-broader potential of information from airborne sensors and the Landsat, SPOT, and similar satellite systems, and with the increased understanding of the value of data from such systems, the time may have come to consider an international Earth resources information service.

Commercial and governmental uses for the information from an Earth resources system have been well demonstrated, particularly in agriculture. For example, estimates of crop growth and yield potentials based on data gathered by Landsat give more accurate and far more timely predictions than do traditional means—a fact of great human and commercial significance.

The private sector has made significant investments in satellite remote sensing products and services. "Value added" weather information, for example, which is information made more useful than raw data, is routinely marketed to weather-dependent businesses and to the media. The petroleum and mineral industries have become insatiable users of SPOT and Landsat data. This foundation for government and private sector applications, plus the broad policy benefits, strongly suggests the ultimate viability of a commercially based space system for the gathering, processing, and marketing of Earth resources information.

As with communications, the potential advantages in the use of Earth resources information lie in both the commercial and public sectors. It is my belief that the operational management of an Earth resources information system should be organized in the United States under an investor-owned, regulated corporation, modeled after the successful experience in communications with Comsat. Furthermore, the Earth resources system should be allied internationally with a user controlled organization, modeled after the equally successful experience with the 110-nation Intelsat and its offshoot, Inmarsat.

In the 1950s and 1960s, while NASA and the private sector pursued the development of communications satellites, the military of both the United States and the U.S.S.R. developed increasingly sophisticated satellites for defense communications, weather forecasting, and intelligence gathering. Most such development was, however, shrouded in secrecy, so the technologies created only indirectly benefited the development of nonmilitary space applications. More rapid transfer of technology from defense to the private sector may be possible in the future.

Among the developing nations, direct broadcast satellites supplying educational programming to widely separated towns and villages offer tremendous opportunities for growth. NASA's 1975–1976 cooperative experiments with India and Brazil with the *ATS-6* direct broadcast satellite, for example, represented the beginning of what may be their most affordable means of access to educational, health, weather, and news services. Although government control of such services can be and has been abused, alternative and more objective regulatory systems can be devised and required as a condition for private or international assistance.

Clearly, then, the age of information has arrived. It is estimated that well over 50 percent of America's gross national product already depends on information-related hardware and services. Further, we rely on satellites for defense and for instant interaction with economies both domestic and beyond. Therefore, the infrastructure of the information age, composed of Intelsat, Inmarsat, et al., however primitive, now forms the rudimentary but functioning central nervous system for the planet Earth, and this unique technological environment offers unlimited potential for the third millennium.

The Settlement of Space: Human Adaptation

(Interplanetary Press)—Apollo Settlements. We watched the first solar starship, *Galaxy*, swing by today. Although *Galaxy's* speed is still a long way from REL 1, its acceleration since its departure from Earth orbit swept it by so fast that it was visible for only about a minute. The crew's report indicated good spirits and confidence. Unlike the excitement that accompanied a historic "first" mission of the past, the mission of *Galaxy* has stirred a deep warmth and contentment in the inhabitants of these settlements. Pride in humankind has at least temporarily unified this planet. More than one viewer has remarked that now we are the UFOs.

From an Interplanetary Press wire story,
June 11, 2029

The exploration of the Moon by the Apollo astronauts became possible because of the physiological adaptability of humans, the technological extension of that ability, and the insights of the then-fledgling space biomedical community. Before crews ventured into space, great uncertainty existed in NASA and elsewhere about the ability of human beings to adapt to the weightless environment of space. The prevailing medical opinion in the early years of the space age stated that humans might not survive exposure to long periods of weightlessness.

To the credit of people such as Dr. Randolph Lovelace and Dr. Charles Berry, NASA became cautious but confident that humans could adapt physically to this new environment. Unfortunately, once it was clear that survival was not an issue, the United States as well as the U.S.S.R. undertook little systematic and scientifically credible and repeatable research into the process of human adaptation to space. The systematic look at exposure to a few months of weightlessness that was performed during the Skylab missions and occasional, well-formulated but severely bounded investigations on Spacelab and Mir are the handful of exceptions that highlight the deficiencies.

Based on the largely qualitative and anecdotal information gained to date, the most common symptoms of space adaptation syndrome include a fullness of the head, headache, and nausea. The aggravating effect of motion, particularly pitch motion—nodding or somersaulting—on the symptoms of space adaptation is clear, whatever may be the underlying physiological explanations.

Unlike the early Soviet cosmonauts, who were in relatively large spacecraft and frequently showed adaptation symptoms, U.S. astronauts in the small Mercury and Gemini spacecraft did not experience the effects of motion because they could not move around. However, with the larger Apollo spacecraft, followed by the even larger Skylab and the space shuttle, astronauts began to experience the same problems that had bothered the early cosmonauts.

About 75 percent of the astronauts who undertake significant motion in weightlessness apparently experience space adaptation symptoms beyond just fullness in the head. Forty percent of the astronauts appear to have strong to severe symptoms, including strong stomach awareness, headache, malaise, and vomiting. However, among about 2 percent of the astronauts, these symptoms disappear within three or four days, and no major headache or stomach symptoms occur subsequently.

Long duration space flights, such as the Skylab, Salyut, and Mir space station missions, confirmed indications from Apollo flights that the adaptation process included much more than multisensory conflicts resembling those of terrestrial motion sickness. Among other symptoms, spot checks disclosed such significant changes as decreases in blood volume and red cell mass, bone demineralization, and muscle atrophy. Thus, the multisensory conflicts that cause symptoms of motion sickness may represent only aggravating factors but not the primary cause of space adaptation syndrome. Systemic—or total body—hemodynamic changes, autonomic nervous system reactions, and interactions between various changes may be far more fundamental.

Not surprisingly, after over two decades of research focused largely on vestibular functions, which give the body a sense of balance, there has been no more progress in eliminating or reducing the adverse effects of space adaptation than there has been in understanding its physiological origins.

A scientific method of developing space medicine would require: a) a systematic protocol to characterize space adaptation and re-adaptation; b) pre-flight, in flight, and post-flight application of that protocol to a cadre of test subjects and observing physicians; and c) repeated flights for that cadre to investigate countermeasures. The application of this protocol should begin by flying physician-subject pairs of the cadre on the relatively short space shuttle flights, then flying the cadre again on longer shuttle missions and eventually on space station *Freedom.*

Without this approach, very conservative design parameters—in themselves potentially detrimental—and much delay will be imposed on the early years of the next great human adventure—the settlement of Mars. Although the Soviet experience with long duration flight seems to be telling us that no insurmountable physiological problems exist, neither they nor we have committed to gathering the data necessary to optimize missions to Mars, to provide appropriate in-flight medical care, and to hurry the time when the settlement of space away from Earth can begin.

The Need To Move Ahead

I am proposing a long range continuing commitment. First, for the coming decade...space station *Freedom,* our critical next step in all our space endeavors. And next, for the new century, back to the Moon. Back to the future. And this time, back to stay. And then a journey into tomorrow, a journey to another planet: a manned mission to Mars....

There are many reasons to explore the universe, but ten very special reasons why America must never stop seeking distant frontiers: the ten courageous astronauts who made the ultimate sacrifice to further the cause of space exploration. They have taken their place in the heavens so that America can take its place in the stars.

President George Bush, July 20, 1989,
on the 20th anniversary
of the first Moon landing

Space is a new frontier and a new challenge for humankind. The nations that effectively utilize technology to exploit the economic, military, and political advantages of space will dominate human activities on this planet well into the next century, if not indefinitely. These nations will also provide the template for the social and political evolution of civilization far into the third millennium.

We are already heavily dependent on space assets for military and civilian communications and meteorological systems. Reliance on space for navigation and reconnaissance increases every day. Industry is now investigating space for new, more efficient manufacturing of materials. Spin-offs from space technology permeate our health care systems, our means of transportation, the computers that manage our lives for better or worse, our entertainment choices, and even the ways we motivate our children to achieve. Every individual in the world is now directly or indirectly affected by activities in space or technologies derived from space.

Humankind thus has the technology to use the advantages and resources of space. The question remains whether we will have the will to embark on this new and indefinite adventure into the future.

One opinion in the United States today argues that there is no hurry. "Space will always be there, and meanwhile, we have more pressing near-term interests here on Earth. Space can be explored scientifically with robots at much lower cost." Unfortunately, however, the challenge of space can no longer be viewed as a mere scientific challenge, as valuable as the science to be done will be. The challenge has become to lead both the human settlement of space and the environmental preservation of our home planet.

Why the hurry? The answer is in today's children. *They* will settle the Moon and then Mars, and they will do it simply because they want to "be there." "Being there" remains the prime human ingredient in all of life's meaningful experiences. The frontier of space has produced a level of excitement and motivation among today's youngsters that has not been seen for nearly a century. Nothing motivates young minds as an unexplored frontier offering the opportunity to "be there." The desire to "be there" will drive today's young away from the established paths of history on a now too confining Earth. Video images and data streams from robots on Mars, no matter how good or complete, will never be enough for today's young, who will settle the Moon and parent the generation that will most likely settle Mars. These settlers of the future, some alive today all over the world and others to come, pose the most critical question of national will we face today.

Perhaps the most important reason for hurrying into space is the absolute moral and political imperative to provide an ever-improving quality of life to the ever-expanding population of Earth. At least ten billion humans are expected to populate Earth by the end of the twenty-first century, yet we currently do not have the technical means to promise them an ever-improving quality of life without increasing our dependence on fossil and nuclear fuels and accepting the adverse technical, environmental, and political consequences of their use.

The environmental hazards of consuming fossil fuels, for example, have caused increasing concern in that the consequences possibly include rising levels of carbon dioxide and methane leading to global warming and the depletion of the protective ozone layer. Natural biological processes also appear to be producing increasing amounts of methane, and there is no conclusive evidence to suggest that human activities rather than natural processes are changing the atmosphere's chemistry. Nevertheless, it is only prudent to seek technological alternatives that minimize the human contribution to the deterioration of our environment.

One of the alternatives to fossil or nuclear fuels may exist on the Moon, in lunar soil. The lunar soil has solar wind gases absorbed in it. One of these gases is Helium-3, an isotope of helium. Helium-3's energy equivalent value relative to coal is about $2 billion per tonne. Using Helium-3 in future fusion power plants will produce electrical energy at about twice the efficiency of today's plants—*without* producing the atmospheric contaminants of coal or the long-lived radioactive wastes of nuclear fission or tritium fusion plants. Power plants fueled by Helium-3 from the Moon could thus supply the environmentally benign and potentially low-cost electrical energy we will need to continually enhance our quality of life and move towards the stars.

Further, on the Moon itself, the processes that would extract Helium-3 from lunar soil would also produce large amounts of hydrogen, carbon compounds, water, and nitrogen. For example, for each ton of Helium-3, 6,100 tons of hydrogen, 5,200 tons of carbon compounds, 3,300 tons of water, and 500 tons of nitrogen would be produced. These by-products would potentially become the consumables necessary to support a permanent lunar mining and research outpost, from where the settlement of Mars could be initiated.

The Future

What will it take for us to do all the above? In the past, what has it taken for Americans to build railroads across a continent or canals between oceans or walk on the Moon?

In separate thinking about this question in 1975, Neil Armstrong and I concluded that it takes a coincidence of four conditions, or in Neil's view, the simultaneous peaking of four cycles of American life. First, a base of technology must exist from which an endeavor can be launched. Second, a period of national uneasiness about America's place in the scheme of human activities must exist. Third, some catalytic event must occur that focuses the national attention on the direction in which to proceed. And fourth, and most important, an articulate and wise leader must sense the first three conditions and put forth with words and action the great things to be accomplished.

Such conditions now exist. Americans now stand on the threshold of the next great adventure of humankind—the settlement of the Moon and Mars. Education, science, civilization, and evolution will all participate in this adventure, as will the purely human desire to "be there."

We have led the first wave of human activities in space—in the face of great risk of failure, but with characteristic confidence in success and in the values of civilization. Most of the conditions necessary for us to undertake the next great adventure are already in place. Our technological and economic foundations are far more solid now than they were in the 1960s, when we decided to go to the Moon. The economic and technical challenges to U.S. world leadership now loom far larger than they did when President Kennedy set a course to the Moon.

President Bush has now provided a vision for the U.S. as a spacefaring nation. The final emotional and legislative energy to re-ignite the torch for space handed to us by President Kennedy can be supplied to generations now alive on Earth by the vision of the protection of Earth's environment and the human settlement of Mars.

2

Chronology

THIS CHAPTER LISTS THE KEY EVENTS in space exploration. The events meet the following selection criteria: (a) significant events on Earth that paved the way to space exploration, or (b) significant first-time events in space. The following nomenclature has been used for space vehicles:

> *Satellite.* An unmanned space vehicle in Earth orbit, the orbit extending up to the geosynchronous orbit.
>
> *Space probe.* An unmanned space vehicle bound for outer space, beyond the geosynchronous orbit.
>
> *Spacecraft.* A manned or inhabited space vehicle.

1232   Rockets were invented and used by the Chinese. This is the earliest known date when humankind used rockets.

1903   Konstantin Eduardovich Tsiolkovsky, a Russian schoolteacher, published the first scientific paper ever written on rocketry. The paper was titled *Research into Interplanetary Space by Means of Rocket Power,* and it discussed the use of liquid fuels for rocket propulsion.

1926   **March 16:** Robert Hutchings Goddard, a U.S. scientist, launched the world's first liquid-fuel rocket at Auburn, Massachusetts, thus providing the first evidence of the feasibility of liquid-fuel propulsion.

1953   Sam Fred Singer, a U.S. engineer, developed the first design for an artificial satellite, called MOUSE, for minimum orbital unmanned satellite of earth.

1957   **October 4:** Humankind's exploration of space began. The U.S.S.R. initiated the exploration by launching into Earth orbit humankind's first satellite, *Sputnik 1*. *Sputnik 1* was a metallic ball weighing 84 kilograms (184 pounds). The average altitude of its orbit was 480 kilometers (300 miles). It remained in orbit for 96 days, during which it orbited Earth 1,400 times.

November 3: The U.S.S.R. launched humankind's first inhabited spacecraft. Called *Sputnik 2*, it was inhabited by a dog named Laika. Laika survived for a week in space. *Sputnik 2*, however, remained in orbit for 103 days and 2,370 orbits.

1958   **February 1:** The United States launched into Earth orbit its first satellite, *Explorer 1*, which commenced the U.S. exploration of space. *Explorer 1* confirmed the existence of Earth's inner radiation belt, the Van Allen belt, located at an altitude of 960 kilometers (600 miles).

March 17: The United States launched *Vanguard 1*, the satellite that provided the first real proof that the planet Earth is pear-shaped, as suggested in 1687 by Sir Isaac Newton. *Vanguard 1* also became the first satellite to use solar power derived from photovoltaic cells.

October 1: The United States formally established the National Aeronautics and Space Administration (NASA).

October 11: The United States launched *Pioneer 1*, the first space probe to leave the vicinity of Earth. The 38-kilogram (84-pound) probe flew 113,120 kilometers (70,700 miles) into space, discovering en route the extent of Earth's Van Allen radiation belt and that radiation exists in bands.

December 6: Humankind made its first attempt to reach the Moon. The United States launched *Pioneer 3* for the Moon, but the space probe reached only 101,728 kilometers (63,580 miles) out, then fell back toward Earth, re-entered the atmosphere, and burned up over Africa.

December 18: The United States launched *SCORE* (Signal Corps Orbiting Relay Experiment), the satellite that transmitted voice communications from space for the first time.

1959   **January 2:** The U.S.S.R. launched *Luna 1,* the first space probe that reached Earth's escape velocity and headed for outer space. *Luna 1* also became the first artificial satellite of the Sun. Radio contact with *Luna 1* was lost at about 594,000 kilometers (371,000 miles).

             August 7: The United States launched *Explorer 6,* the satellite that discovered Earth's electrical current system.

             September 12: The U.S.S.R. launched *Luna 2,* the first space probe that reached another body in space—the Moon. *Luna 2* crashed into the Moon's Sea of Serenity.

             October 4: The U.S.S.R. launched *Luna 3,* the space probe that gave humankind the first glimpses of the far side of the Moon. *Luna 3*'s orbit was a giant ellipse around Earth. The orbit's apogee was nearly 480,000 kilometers (300,000 miles), so it encompassed the Moon, thus sending *Luna 3* behind the Moon, from where it photographed the Moon's dark side.

1960   **April 1:** The United States launched *TIROS 1* (Television and Infrared Observation Satellite), the world's first weather satellite.

             April 13: The United States launched *Transit 1B,* the world's first navigation satellite.

             May 24: The United States launched *Midas 2,* the world's first infrared surveillance satellite.

             August 12: The United States launched *Echo 1,* the world's first communications satellite. *Echo 1* was a 30-meter (100-foot) inflatable, aluminized-mylar sphere capable of reflecting radio waves back down to Earth.

             August 20: The first living beings launched into Earth orbit returned back to Earth and were recovered alive. The two dogs named Belka and Strelka, six mice, and some insects had been launched by the U.S.S.R. aboard *Sputnik 5* on August 19, 1960.

1961   **February 12:** Humankind made its first attempt to reach Venus. The U.S.S.R. launched *Venera 1,* using *Sputnik 8,* which was in Earth orbit, as a launch platform. *Venera 1* flew past Venus.

             April 12: A human being flew in space for the first time. Yuri A. Gagarin of the U.S.S.R. piloted the spacecraft *Vostok 1* into Earth orbit. Gagarin completed one orbit.

1961
cont.
May 25: U.S. President John F. Kennedy, in an address to a joint session of the U.S. Congress, established a lunar goal for the United States and committed the nation to achieving it. President Kennedy said, "I believe this nation should commit itself to achieving the goal, before this decade is out, of landing a man on the Moon and returning him safely to Earth."

1962
February 20: The United States embarked on manned space exploration by launching its first man into orbit. Astronaut John Glenn piloted *Mercury-Friendship 7* into space for three orbits.

April 23: The United States launched *Ranger 4*, the first U.S. space probe to reach the Moon. *Ranger 4* impacted the Moon.

July 10: The United States launched *Telstar 1*, the first commercial satellite in space. *Telstar 1* was a communication satellite of the American Telephone and Telegraph company.

July 23: The first telecast from the United States was relayed across the Atlantic by the U.S. satellite *Telstar 1*, launched into orbit on July 10, 1962.

August 11 & 12: The U.S.S.R. launched a cosmonaut on each date, thus putting two men in orbit simultaneously for the first time. The two cosmonauts were Andrian Nikolayev in *Vostok 3* and Pavel Popovich in *Vostok 4*.

December 14: The U.S. space probe *Mariner 2*, launched on August 27, 1962, flew within 34,637 kilometers (21,648 miles) of Venus. *Mariner 2* confirmed that the solar wind comprises a steady stream of protons and electrons. This provided the first concrete evidence that interplanetary space is filled with a fully ionized hot plasma of solar origin.

1963
May 15: The United States concluded its one-man space flight program with the launch of Gordon Cooper in *Mercury 8*. Cooper completed 22 orbits, and since then, no American has been launched alone into space.

June 16: The U.S.S.R. launched Valentina Tereshkova, the first woman in space, aboard *Vostok 6*. She completed 48 orbits in 70 hours and 50 minutes.

July 26: The United States launched *Syncom 2*, humankind's first satellite to achieve geosynchronous orbit. It achieved the orbit in mid-August over Brazil.

1963
cont.
November 27: The use of liquid hydrogen as a rocket fuel was demonstrated for the first time by the *Centaur 2* rocket of the United States.

December 21: The United States launched *TIROS 8* (Television and Infrared Observation Satellite), the first satellite capable of automatic picture transmission.

1964
August 28: The United States launched *Nimbus 1*, the first satellite to take infrared pictures of Earth.

October 12: The U.S.S.R. launched the first three-man crew into orbit aboard *Voskhod 1*. The crew members were Vladimir Komarov, Boris Yegorov, and Konstantin Feoktistov.

1965
March 18: A human walked in space for the first time. Alexei Leonov of the U.S.S.R. climbed out of the spacecraft *Voskhod 2* and became the first human to walk in space. The 10-minute walk was tethered: Leonov was attached to *Voskhod 2* by an umbilical cord.

March 24: *Ranger 9* of the United States sent to Earth the first live television pictures of the Moon before it impacted the lunar surface. *Ranger 9* was launched on March 21, 1965.

April 6: The first private-industry geosynchronous satellite, *Intelsat 1*, also called *Early Bird,* was launched into orbit. Communications Satellite Corporation developed the satellite.

June 3: Ed White became the first U.S. astronaut and the second human in history to walk in space. During the walk, White became the first person to use a hand-held maneuvering unit to propel himself in space. White's walk was tethered, and his spacecraft was *Gemini 4,* launched on the same date as the walk. White walked in space for 21 minutes.

July 14: *Mariner 4,* of the United States, came within 9,790 kilometers (6,118 miles) of Mars and radioed back the first close-up pictures of the planet. *Mariner 4* was launched on November 28, 1964. It traveled 520 million kilometers (325 million miles) for the Mars encounter and the 21 pictures of Mars that it radioed back to Earth.

August 21: The United States launched *Gemini 5,* the first spacecraft on which fuel cells were used.

1965
cont.
November 26: France became the third nation to launch a satellite independently. The name of the satellite was *A1 Asterix.*

December 15: The United States achieved the first rendezvous of two spacecraft in Earth orbit. *Gemini 6,* manned by Walter Schirra and Thomas Stafford, and *Gemini 7,* manned by Frank Borman and James Lovell, rendezvoused to within three feet of each other. *Gemini 7* had been launched first, on December 4, 1965; *Gemini 6* was launched on December 15, 1965.

1966
February 3: Humankind achieved its first soft landing on another body in space. *Luna 9,* an unmanned space probe of the U.S.S.R., launched on January 31, 1966, made a soft landing in the Moon's Ocean of Storms.

March 1: *Venera 3,* of the U.S.S.R., entered the atmosphere of Venus, thus becoming humankind's first space probe to enter another planet's atmosphere. During its approach to Venus, *Venera 3* radioed back valuable data on the planet, but communications broke down after the spacecraft entered Venus's atmosphere. *Venera 3* was launched on November 16, 1965.

March 16: Two space vehicles docked in space for the first time. *Gemini 8* of the United States, launched on the same date, docked with an Agena target vehicle, an unmanned vehicle for practicing docking in space.

June 2: *Surveyor 1* became the first U.S. space probe, and humankind's second, to make a soft landing on another body in space. *Surveyor 1,* launched on May 30, 1966, landed in the Moon's Ocean of Storms.

September 12: Man created artificial gravity in space for the first time when the U.S. spacecraft *Gemini 11,* tied to an Agena target vehicle, spun in space with a slow spin. Astronaut Charles Conrad reported "a little artificial gravity" created by the spin. A spin creates centrifugal force, which in space is artificial gravity.

November 11: The United States launched *Gemini 12,* the mission during which the astronaut Edwin Aldrin, Jr., set the record for the longest space walk yet: five and one-half hours.

December 8: The United States and the U.S.S.R. announced an agreement on a treaty to place no bombs in orbit. The treaty was drafted by the United Nations Committee on the Peaceful Uses

1966
cont.
of Outer Space and formally signed on January 27, 1967. The U.S. Senate ratified the treaty on April 25, 1967; the Presidium of the Supreme Soviet ratified it on May 19, 1967.

1967
January 27: The U.S. space program suffered its first casualties. Astronauts Virgil "Gus" Grissom, Ed White, and Roger Chaffee were practicing for their planned February 21 launch of *Apollo 1* when a fire consumed the spacecraft. A review board later determined that the command module's atmosphere had been 100 percent oxygen. The oxygen ignited, killing all three astronauts.

April 24: The U.S.S.R. space program suffered its first casualty. Cosmonaut Vladimir Komarov was piloting Soyuz 1 back down to Earth when the spacecraft's parachute harness got entangled. Soyuz 1 crashed, killing Komarov.

November 29: Australia became the world's fourth space nation by launching its first satellite, *WRESAT* (Weapons Research Establishment Satellite).

1968
December 24: Human beings orbited another body in space for the first time. The U.S. spacecraft *Apollo 8,* launched from Earth on December 21, 1968, went into orbit around the Moon. *Apollo 8* was manned by Frank Borman, James Lovell, and William Anders.

1969
July 20: At 10:56 P.M. U.S. eastern daylight time, Neil Armstrong became the first human being to set foot on another celestial body. The body was the Moon, and the spacecraft was the U.S. *Apollo 11,* launched from Earth on July 16, 1969. The crew was Neil Armstrong, commander; Edwin Aldrin, Jr., lunar module pilot; and Michael Collins, command module pilot. On July 20, while Collins orbited overhead in the command module, Armstrong and Aldrin piloted the lunar module down to the Moon's Sea of Tranquility and spent 21 hours and 37 minutes on the lunar surface.

1970
February 11: Japan became the fifth nation to launch a satellite into orbit. The satellite was named *Osumi.*

April 13: The first life-threatening mishap occurred in space. *Apollo 13* of the United States, launched on April 11, 1970, was headed for the Moon when one of its liquid oxygen tanks exploded. Having lost most of its fuel, *Apollo 13* could not land on the Moon. The crew's ingenuity in using the Moon's gravity

1970
cont.
as a slingshot to return the spacecraft to Earth saved the crew: James Lovell, John Swigert, and Fred Haise.

April 24: China became the sixth nation to launch a satellite, *China 1,* into orbit. The satellite remained in orbit for two months, broadcasting the anthem "The East is Red."

September 21: For the first time, an unmanned space probe returned to Earth with soil samples from another body in space. *Luna 16* of the U.S.S.R., launched on September 12, 1970, landed on the Moon, collected 10 kilograms (22 pounds) of lunar soil, and returned to Earth.

December 15: Humankind achieved its first soft landing on another planet. The Soviet space probe *Venera 7,* in orbit around Venus, ejected a capsule that landed on Venus and transmitted data for 23 minutes. This was also the first time a transmission was sent to Earth from the surface of another planet. *Venera 7* was launched from Earth on August 17, 1970.

1971
April 19: The U.S.S.R. launched *Salyut 1,* which became humankind's first space station in Earth orbit. It was manned for 22 days by Georgi Dobrovolsky, Viktor Patsayev, and Vladislav Volkov, who were launched aboard *Soyuz 11* on June 6, 1971. On June 29, 1971, the three cosmonauts were killed while returning to Earth. During re-entry, a valve opened accidentally, and the spacecraft lost its cabin pressure.

November 27: Humankind achieved its first soft landing on its nearest neighbor, Mars. *Mars 2* of the U.S.S.R., launched on May 19, 1971, ejected a lander while orbiting Mars. The lander made a soft landing and briefly transmitted data from the planet's surface.

1972
March 2: The United States launched humankind's first interstellar space probe, *Pioneer 10. Pioneer 10*'s mission was to fly by Jupiter and Saturn, then plunge on beyond the Solar System into the Milky Way Galaxy, headed for the stars. At this very moment, it is hurtling on into deep space.

December 13: Humankind's last manned expedition to the Moon as yet lifted off the lunar surface at 5:55 P.M. U.S. eastern standard time. The spacecraft was *Apollo 17* of the United States, launched from Earth on December 7, 1972. *Apollo 17* returned to Earth on December 19, 1972, 310 hours and 51 minutes after launch, thus making it the longest manned lunar expedition to date.

1973 **May 14:** The United States launched its first space station, *Skylab*. In the same year, *Skylab* was host to three different three-man crews before it re-entered the atmosphere on July 11, 1979.

November 3: The United States launched *Mariner 10*, humankind's first interplanetary space probe that used one planet's gravity to hurl itself to another planet. *Mariner 10* flew past Venus and went on to Mercury.

December 3: Humankind got its first close look at Jupiter. The U.S. space probe *Pioneer 10*, launched on March 2, 1972, flew past the largest planet of our Solar System and radioed to Earth the first close-up pictures of the planet.

1974 **February 8:** The longest U.S. space mission to date ended. Astronauts Gerald Carr, Edward Gibson, and William Pogue ended their stay aboard the first and so far the only U.S. space station, *Skylab*. Their mission lasted 2,017 hours and 16 minutes.

March 26: The U.S.S.R. launched *Kosmos 637*, the first Soviet satellite to achieve geosynchronous orbit.

March 29: The planet Mercury got its closest scrutiny yet from humankind when the U.S. space probe *Mariner 10* flew less than 800 kilometers (500 miles) from the planet. *Mariner 10*, launched from Earth on November 3, 1973, flew past Venus, then went into a solar orbit in which it passed Mercury about once every six months. *Mariner 10* passed Mercury twice more after this first pass: on September 21, 1974, and on March 16, 1975. It then ran out of attitude-control fuel and lost radio contact with Earth.

1975 **July 17:** For the first time, spacecraft of two nations docked together in Earth orbit, and the first international handshake in space occurred. *Apollo 18* of the United States docked with *Soyuz 19* of the U.S.S.R. In the U.S. spacecraft were Donald Slayton, Vance Brand, and Thomas Stafford. In the Soviet spacecraft were Alexei Leonov and Valeri Kubasov.

October 22: Pictures were transmitted for the first time from the surface of another planet. They were transmitted by *Venera 9* of the U.S.S.R., launched on June 8, 1975. The pictures were also humankind's first look at the surface of Venus, which is usually clouded over. *Venera 9* functioned for 53 minutes before the high atmospheric pressure of Venus crushed it.

1976   **February 10:** Humankind got its first close look at Saturn. *Pioneer 10,* the U.S. space probe launched on March 2, 1972, reached Saturn and transmitted close-up pictures of the planet.

July 20: Pictures were transmitted for the first time from the surface of Mars when the U.S. space probe *Viking 1*'s lander touched down on Martian soil. *Viking 1* was launched from Earth on August 20, 1975. It achieved Mars orbit on June 19, 1976.

August 9: The U.S.S.R. launched *Luna 24,* humankind's last lunar mission to date. *Luna 24* brought to an end the era of lunar exploration that had started in 1959 when *Luna 3* first photographed the dark side of the Moon. Between 1959 and 1976, the U.S.S.R. soft-landed seven unmanned space probes, while the United States conducted four unmanned and six manned landings on the Moon.

1978   **January 16:** A crew exchanged their spacecraft for another for the first time in space. The crew of the U.S.S.R.'s *Soyuz 27* returned to Earth in *Soyuz 26,* which had been taken up by the crew of the space station *Salyut 6* on December 10, 1977.

January 22: An unmanned space vehicle was used for the first time to ferry supplies to humans already in space. *Progress 1,* an unmanned vehicle of the U.S.S.R. launched on January 20, 1978, took supplies into Earth orbit for the crew manning the Soviet space station *Salyut 6.*

March 2: Vladimir Remek of Czechoslovakia became the first citizen of a country other than the United States or the U.S.S.R. to orbit Earth. Remek was a guest aboard the U.S.S.R.'s *Soyuz 28.*

October 1: U.S. President Jimmy Carter awarded the first Congressional Space Medals of Honor to six former astronauts: Neil Armstrong, Frank Borman, Charles Conrad, John Glenn, Virgil "Gus" Grissom (posthumously), and Alan Shepard.

1979   **March 5:** The third U.S. interstellar space probe, *Voyager 1,* launched from Earth on September 5, 1977, arrived at Jupiter and gave humankind its first close-up look at the planet's great red spot and the planet's moons.

1980   **November 12:** The U.S. interstellar space probe *Voyager 1,* launched from Earth on September 5, 1977, passed within 123,200 kilometers (77,000 miles) of Saturn and was redirected

1980
cont. to fly past Saturn's largest moon, Titan, which is the only moon in the Solar System known to have an atmosphere. *Voyager 1* thus provided the first close-up look at Titan.

1981 **March 12:** Victor Savinykh of the U.S.S.R., launched aboard *Soyuz T4,* became the one-hundredth human in space.

April 12: The United States launched the world's first reusable spacecraft—the space shuttle. Named *Columbia,* the 77,206-kilogram (76-ton) shuttle was piloted into orbit by John Young and Robert Crippen for a 54-hour, 22-minute maiden flight of 36 orbits, at the end of which it landed like an aircraft at Edwards Air Force Base, California, thus also becoming the first U.S. manned space flight ever to return to solid ground.

1983 **June 18:** The United States launched its first woman—the world's third—into Earth orbit. She was Sally K. Ride, a mission specialist aboard the space shuttle *Challenger,* mission STS 7.

1984 **February 8:** Astronaut Bruce McCandless of the United States performed the world's first untethered space walk. The mission was STS 41B, launched on February 3, 1984. The walk was performed to test the manned maneuvering unit, a device for astronauts to maneuver in space during a walk. McCandless went about 90 meters (300 feet) away from his spacecraft, untethered all the time. The spacecraft was the shuttle *Challenger.*

October 2: The crew of the U.S.S.R. space station *Salyut 7* established the record for the longest stay yet in Earth orbit—237 days. The crew was Oleg Atkov, Leonid Kizim, and Vladimir Solovyev.

1985 **October 30:** The United States launched the largest crew ever launched to date by any nation aboard a single spacecraft. The eight-member crew was launched aboard the space shuttle *Challenger,* mission STS 61A.

December 3: The United States launched the space shuttle *Atlantis.* The launch established a new record for the number of spacecraft—nine—launched into Earth orbit by one nation in a single year.

1986 **January 24:** Humankind's fourth interstellar space probe, *Voyager 2* of the United States, launched from Earth on August 20, 1977, flew by Uranus, thus giving humankind its first close-up look at the planet. In the process, *Voyager 2* discovered ten new moons of Uranus.

1986   **January 28:** As yet the worst disaster of space exploration oc-
cont.   curred when the U.S. space shuttle *Challenger* exploded 73
             seconds after lift-off, killing all seven of its crew. The crew were
             Francis Scobee, Michael Smith, Judith Resnik, Ellison Onizuka,
             Ronald McNair, Gregory Jarvis, and Sharon Christa McAuliffe.

March 9: The Soviet space probe *Vega 2,* launched from Earth
on December 21, 1984, flew within 7,979 kilometers (4,987
miles) of Halley's Comet and radioed to Earth the first pictures
ever seen of the comet's core.

March 14: The spacecraft *Giotto* of the European Space Agency
made humankind's closest ever approach to Halley's Comet—
586 kilometers (366 miles).

July 16: Leonid Kizim of the U.S.S.R. established the still-stand-
ing record for the most experienced human in space, having
logged a total of 374 days and 18 hours in Earth orbit.

1989   **August 24:** Humankind's fourth interstellar space probe, *Voyager
             2* of the United States, launched from Earth on August 20, 1977,
             arrived at Neptune and gave humankind its first close-up look
             at the planet. Pictures radioed back to Earth by *Voyager 2* took
             four hours to travel over 2.8 billion miles. *Voyager 2'*s itinerary
             took it past Jupiter, Saturn, Uranus, and Neptune—a trajectory
             that is possible only once every 176 years. Having completed its
             mission in the Solar System, *Voyager 2* is now plunging on into
             the Milky Way Galaxy, headed for the stars.

3

Biographies

THIS CHAPTER CONTAINS BIOGRAPHICAL SKETCHES of the key people who contributed to space exploration, either through research and experimentation or by participating in key missions. Biographies of all those who walked on the Moon are listed, as are biographies of those who died as space explorers. The biographies are arranged alphabetically by last name.

Edwin Eugene Aldrin, Jr.

U.S. astronaut, born January 20, 1930. An air force colonel, Aldrin graduated from West Point Military Academy in 1951 with a bachelor's degree in mechanical engineering. He joined the air force and flew combat missions in the Korean War, after which he served in West Germany. In 1963, he earned a doctorate degree in astronautics from the Massachusetts Institute of Technology. That same year, he was recruited by NASA as an astronaut. Aldrin's first flight into space was aboard *Gemini 12*, during which he set the record of the longest space walk yet—five and one-half hours. His next flight took him to the Moon as the lunar module pilot on the historic *Apollo 11*, the first manned lunar landing mission. Aldrin thus became the second human to set foot on another celestial body, the first being his commander on the same mission, Neil Armstrong. Aldrin retired from NASA in 1971.

William Alison Anders

U.S. astronaut, born October 17, 1933. An air force major, Anders was a fighter pilot in the Air Defense Command and holds a master's degree in nuclear engineering. He also served as a technical manager at the Air Force Weapons Laboratory. He was recruited into NASA in 1963. Anders's only flight into space was aboard *Apollo 8,* when he became one of the first three humans to leave Earth for another celestial body. *Apollo 8* was humankind's first manned flight to the Moon. The mission took Anders, Frank Borman, and James Lovell for ten lunar orbits on Christmas Day, 1968. Anders resigned from the space program after *Apollo 8.*

Neil Alden Armstrong

U.S. astronaut, born August 5, 1930. A civilian pilot, Armstrong was the first human to set foot on another celestial body—the Moon. Armstrong earned a pilot's license at age 16, then became a naval air cadet in 1947. He enrolled at Purdue University, where he planned to earn a bachelor's degree in aeronautical engineering, but his studies were interrupted in 1950 by the Korean War. During the war, Armstrong was shot down once and earned three Air Medals. In 1955, he became a research pilot for the NASA precursor called the National Advisory Committee for Aeronautics. By the time he was recruited as an astronaut in 1962, he had logged over 1,100 hours of flying time in jets, including the X-15 rocket plane. Armstrong's first flight into space was in 1966 aboard *Gemini 8.* His next space flight was aboard the historic *Apollo 11,* which he commanded to the Moon as the first manned lunar landing mission. It was during this mission that Armstrong became the first human to set foot on another celestial body. Armstrong resigned from NASA in 1971. For his achievements in space, Armstrong was awarded the Congressional Space Medal of Honor on October 1, 1978, by President Jimmy Carter.

Alan LaVern Bean

U.S. astronaut, born March 15, 1932. A navy captain, Bean earned a bachelor's degree in aeronautical engineering from the University of Texas in 1955. He then joined the navy and trained and served as a test pilot. NASA recruited Bean in 1963 and, in 1969, sent him to the Moon as the lunar module pilot on *Apollo 12,* the second manned lunar landing mission. Bean thus became the fourth human to set foot on another celestial body. Bean went back into space in 1973, this time in Earth orbit, as commander of the *Skylab 3* mission. Bean retired from the navy in 1975 but stayed on at NASA for a while. He now devotes his time to painting.

Frank Borman

U.S. astronaut, born March 14, 1928. An air force colonel, Borman graduated from West Point Military Academy in 1950. He was then commissioned in the air force and served in the 44th Fighter Bomber Squadron in the Philippines from 1951–1956. After that he taught at the Air Force Fighter Weapons School. In 1957, he earned a master's degree in aeronautical engineering from the California Institute of Technology. He then returned to West Point to teach. He also taught at the Air Force Aerospace Research Pilots School. NASA recruited Borman in 1962. His first flight into space was aboard *Gemini 7*, which along with *Gemini 6* made history by executing the first rendezvous between two spacecraft in space. Borman's next flight, *Apollo 8*, was even more historic, for he became one of the first three humans to leave Earth for another celestial body—the Moon. *Apollo 8* completed ten lunar orbits on Christmas Day, 1968, thus paving the way for a lunar landing the following year. Borman resigned from the space program in 1970.

Malcolm Scott Carpenter

U.S. astronaut, born May 1, 1925. Carpenter was one of the first seven U.S. astronauts. A navy commander, he graduated from the University of Colorado with a degree in aeronautical engineering. He began his flight training during World War II but couldn't complete it before the war ended. After the war he went to college, then rejoined the navy in 1949, completed his flight training, and flew antisubmarine missions in the Yellow Sea, the South China Sea, and the Formosa Straits. In 1954, Carpenter attended the navy's test pilot school, and in 1958, he was assigned to the U.S. aircraft carrier *Hornet*. By the time he was selected as an astronaut, he had logged over 3,000 hours of flying time. As an astronaut, he piloted the *Mercury-Aurora 7* into Earth orbit in 1962.

Eugene Andrew Cernan

U.S. astronaut, born March 14, 1934. A navy captain, Cernan was commissioned into the navy in 1956. He trained as a test pilot and earned a bachelor's degree in electrical engineering from Purdue University and then a master's degree in aeronautical engineering from the U.S. Naval Postgraduate School. NASA recruited Cernan in 1963 during its second round of astronaut selections. Cernan first flew in space in 1966 aboard *Gemini 9*. His next mission was *Apollo 10* in 1969, which was the final precursor to the manned lunar landing mission of *Apollo 11*. Cernan went to the Moon on *Apollo 10*, and as the lunar module's pilot, he maneuvered the module down to within 1,500 meters (5,000 feet) of the lunar surface. In 1972, Cernan went back to the Moon, this time as commander of *Apollo 17*, the last flight to the Moon to date. He thus became the eleventh human to set foot on another celestial body. After *Apollo 17*, Cernan was appointed deputy director of the Apollo-Soyuz project. He resigned from the navy and NASA on July 1, 1976.

Roger Bruce Chaffee

U.S. astronaut, born February 15, 1935, died January 27, 1967. A navy lieutenant commander, Chaffee earned a bachelor's degree in aeronautical engineering from Purdue University in 1957, after which he became a navy pilot. His first application to NASA's astronaut cadre was not successful, but Chaffee tried again in 1963 and succeeded. He was scheduled to fly as communications officer aboard *Apollo 1*, the first of the missions in preparation for the lunar landing. Chaffee and his crew mates, veteran astronauts Virgil Grissom and Ed White, were training in *Apollo 1* on January 27, 1967, when an electrical short circuit ignited the oxygen-rich atmosphere of the spacecraft and killed all three astronauts. This was the first disaster to strike the U.S. space exploration effort.

Michael Collins

U.S. astronaut, born October 31, 1930. An air force major general, Collins graduated from West Point Military Academy in 1952 and joined the air force as a test pilot of experimental jet fighters. NASA recruited him as an astronaut in 1963 and sent him into space as the pilot of *Gemini 10* in 1966. Collins's next flight was to the Moon as the command module pilot on the historic *Apollo 11*, the first manned lunar landing mission. Collins stayed in lunar orbit while his two crew mates landed on the Moon. Collins resigned from NASA after the *Apollo 11* mission.

Charles Conrad, Jr.

U.S. astronaut, born June 2, 1930. A navy captain, Conrad was recruited in 1962 in NASA's second round of astronaut recruitment. He was a graduate of Princeton University, holding a bachelor's degree in aeronautical engineering. In 1953, Conrad enlisted in the navy and became a test pilot and, later, a flight instructor. His first flight into space was in 1965 on *Gemini 5* under the command of Gordon Cooper. In 1966, he flew again, this time as commander of *Gemini 11*. Conrad commanded the spacecraft to a successful docking with a target vehicle, thus proving the feasibility of docking concepts that were to pave the way to a lunar landing. In 1969, Conrad commanded *Apollo 12*, the second manned lunar landing mission to the Moon. On November 19, 1969, Conrad became the third human to set foot on another celestial body. Conrad and his crew mate, Alan Bean, spent a little over 31 hours on the lunar surface. Conrad's final flight into space was aboard *Skylab 2* in 1973, during which he docked with *Skylab 1*, which was already in orbit but damaged. Conrad and his two crew mates then conducted a historic repair walk in space that saved the multimillion-dollar *Skylab* program from potential failure. For this, and for his earlier feats in space, Conrad was awarded the Congressional Space Medal of Honor by President Jimmy Carter on October 1, 1978. He resigned from the astronaut program in 1974 and began a career in private industry as an executive.

Leroy Gordon Cooper, Jr.

U.S. astronaut, born March 6, 1927. Cooper was one of the first seven U.S. astronauts. An air force colonel, he holds a bachelor of science degree from the Air Force Institute of Technology and a doctorate in aeronautical engineering from Oklahoma City University. Cooper served briefly in the Marines and the army, then transferred to the air force in 1949. By the time he was selected as an astronaut, he had logged over 2,500 hours of flying time, over 1,500 of them in high-performance jets. Cooper piloted *Mercury-Faith 7* into orbit in 1963, and that mission was the last time a U.S. astronaut was launched alone into space. Cooper also commanded *Gemini 5* into Earth orbit in 1965.

Robert Laurel Crippen

U.S. astronaut, born September 11, 1937. A navy captain, Crippen was Commander John Young's pilot on the historic maiden voyage of the space shuttle *Columbia*—humankind's first reusable spacecraft—on April 12, 1981. Crippen earned a bachelor's degree in aerospace engineering from the University of Texas at Austin in 1960. In 1966, he entered the Orbiting Laboratory program of the Department of Defense, and NASA recruited him in 1969. In the following years, he became commander of the Skylab Medical Experiments Altitude Test and was a member of the support crew for *Skylab 2, Skylab 3, Skylab 4,* and the Apollo-Soyuz link missions. After the historic maiden voyage of the space shuttle in 1981, Crippen flew the shuttle again, this time as commander, on three more shuttle missions in 1983 and 1984. Crippen is now deputy director of NASA Space Transportation System Operations.

Georgy Timofeyevich Dobrovolsky

Cosmonaut, born June 1, 1928, died June 29, 1971. An air force lieutenant colonel, Dobrovolsky was selected as a cosmonaut in 1963. In 1971, Dobrovolsky commanded *Soyuz 11* and a crew of two others into space and docked with the famous space station *Salyut 1.* The crew remained in space for 24 days, performing meteorological and plant-growth experiments. On June 29, 1971, Dobrovolsky was commanding *Soyuz 11* back down to Earth when a valve in the spacecraft opened accidentally during re-entry and depressurized the cabin. Dobrovolsky and his crew were found dead upon recovery of the craft.

Charles Moss Duke, Jr.

U.S. astronaut, born October 3, 1935. An air force brigadier general, Duke graduated from the U.S. Naval Academy in 1957 with a bachelor's degree in naval sciences. In 1964, he earned a master's degree in aeronautics from the Massachusetts Institute of Technology. He was an instructor in the Air Force Aerospace Research Pilot School when NASA recruited him in 1966. Duke's only flight into space was as the lunar module pilot on the historic *Apollo 16,* the sixth manned lunar landing mission. Duke thus became the tenth human to set foot on another celestial body.

Yuri Alexeyevich Gagarin

Cosmonaut, born March 9, 1934, died March 27, 1968. Born into a carpenter's family, Gagarin will always be remembered as the first human to orbit the planet Earth. During World War II was the first time Gagarin saw an airplane. In 1945, his family moved to Gzhatzk, where Gagarin completed school and then attended a trade-training school. He worked in a foundry for a while, then went to an industrial training college in Saratov, where he joined a flying club and took flight and parachute training. Once he had experienced the thrills of flying, Gagarin never left it. In 1957, he graduated from the air force training school at Orenburg, and in 1960, he was selected as a cosmonaut. By this time, he had logged a mere 230 hours of flying time. Intensive training followed at the cosmonaut training center at Zvezdniy Gorodok, or Star City, and on April 8, 1961, Gagarin was chosen to pilot the *Vostok 1* into space four days later. On April 12, 1961, at 9:07 A.M. Moscow time, Gagarin lifted off from Earth and roared into space, thus becoming the first human in space. He landed back on Earth at 10:55 A.M. Moscow time. This short flight pioneered a new era in the history of humankind. Gagarin was killed in an air crash in 1968. The city of Gzhatzk, where Gagarin completed school, is now called Gagarin, and a 40-meter (130-foot) high titanium obelisk marks the spot where Gagarin landed after his historic solo flight into space.

John Herschel Glenn, Jr.

U.S. astronaut, born July 18, 1921. Glenn was one of the first seven U.S. astronauts. A Marine colonel, he holds a bachelor of science degree in mathematics from Muskingum College. Glenn piloted fighters in the Pacific during World War II and then over North China after the war. He flew combat missions in the Korean War and shot down three MiGs on the Yalu River. He then became a test pilot. In July 1957, he set a record by flying an F8U from Los Angeles to New York in 3 hours and 23 minutes. By the time he was selected as an astronaut, Glenn had won the Distinguished Flying Cross five times and an Air Medal with 18 clusters, and he had logged over 5,000 hours of flying time. Glenn piloted *Mercury-Friendship 7* into Earth orbit in 1962.

Robert Hutchings Goddard

U.S. rocket pioneer, born October 5, 1882, died August 10, 1945. Since his youngest days, Goddard was fascinated by the idea of space travel. On October 19, 1899, at age 17, Goddard sat perched in a cherry tree and dreamed of traveling to Mars; he spent the rest of his life trying to turn that dream into reality. Inspired by the writings of Russian rocket researcher Konstantin Eduardovich Tsiolkovsky, Goddard decided to test Tsiolkovsky's theories on rocket propulsion. After more than two decades of research, experiments, and plenty of failures, on March 16, 1926, at Auburn, Massachusetts, Goddard launched the world's first liquid-fuel rocket. It flew 56 meters (184 feet) high in 2.5 seconds. Goddard launched a few more rockets in Massachusetts, but then the Massachusetts legislature passed laws that forbid Goddard to do any more rocket work in that state, thus forcing him to leave. He resettled in Roswell, New Mexico, where he continued his work. During World War II, he helped the U.S. Army design rockets and rocket launchers, and then he returned to New Mexico and continued his research and experiments till his death. A museum now honors him in Roswell. The museum houses Goddard's workshops and some of his original rocket parts and design sketches and drawings.

Virgil Ivan ("Gus") Grissom

U.S. astronaut, born April 3, 1926, died January 27, 1967. Grissom was one of the first seven U.S. astronauts. An air force lieutenant colonel, Grissom earned a Bachelor of Science degree in mechanical engineering from Purdue University in 1950. He flew combat missions during the Korean War, then became a flight instructor, from which he went on to become a test pilot for new jet fighters at Wright-Patterson Air Force Base. By the time he became an astronaut, he had logged over 3,000 hours of flying time. Grissom piloted *Mercury-Liberty Bell-7* into a suborbital flight in 1961, and in 1965, he commanded *Gemini 3* into Earth orbit. Grissom was then chosen to command the *Apollo 1* mission into space. He was training for the mission on January 27, 1967, when an electrical short circuit ignited the spacecraft's oxygen-rich atmosphere. Grissom and his crew of Roger Chaffee and Ed White died in the fire. Grissom was posthumously honored by President Jimmy Carter with the Congressional Space Medal of Honor on October 1, 1978, making him one of only seven astronauts who have been honored with the medal so far.

James Benson Irwin

U.S. astronaut, born March 17, 1930. An air force colonel, Irwin graduated from the U.S. Naval Academy in 1951 with a bachelor's degree in naval sciences. He became a test pilot. In 1957, he earned a master's degree in aeronautical and instrumentation engineering from the University of Michigan. NASA recruited Irwin in 1966. His first and only flight into space was as the lunar module pilot on the historic *Apollo 15*, the fifth manned lunar landing mission. Irwin thus became the eighth human to set foot on another celestial body.

Vladimir Mikhaylovich Komarov

Cosmonaut, born March 16, 1927, died April 24, 1967. A Soviet air force colonel, Komarov was a born flier. He joined flight training at 15, was trained at Soviet air force schools, and became a pilot in 1949. In 1959, he graduated from the Zhukovsky Air Force Engineering Academy. (Later, he graduated from three other air force colleges as well.) Komarov was chosen for the space program during the first round of cosmonaut selections, but he could not fly immediately because of a minor heart murmur discovered shortly after his selection. In 1964, however, Komarov commanded *Voskhod 1* into space with a crew of two other cosmonauts. This was the first time that three humans flew together into space. Three years later, Komarov was launched again, alone this time, aboard *Soyuz 1*. During the descent back to Earth on April 24, 1967, the lines of the spacecraft's parachute got tangled, and Komarov crashed to his death, becoming the first casualty of the Soviet space exploration effort. Komarov's ashes are buried in the Kremlin Wall.

Sergei Pavlovich Korolyov

Soviet rocket engineer, born December 30, 1907, died January 14, 1966. Korolyov is recognized as the father of the Soviet space program and one of the pioneering fathers of modern rocketry. Not long after he graduated from the Bauman Higher Technical School, in the Ukraine, Korolyov met Konstantin Tsiolkovsky, the pioneer rocketry researcher, who impressed Korolyov with his ideas on space travel and how to make it possible. Meeting Tsiolkovsky was the beginning of Korolyov's dedication to space exploration. In 1933, he founded a now-famous group of rocket engineers and started thinking in earnest about high-powered rockets that would boost humans into space. In 1945, at the end of World War II, Korolyov was assigned to develop advanced versions of the German V-2 rocket missile that had wreaked havoc on England during the war. Over the next few years, Korolyov worked in the Soviet military industries and developed the U.S.S.R.'s first ballistic missile. Once the missile had been successfully tested, Korolyov went back to concentrating on how to get humans into space. He developed the Sputnik program, which led to the manned Vostok, Voskhod, and Soyuz programs. Korolyov also developed the R7 missile, which is known in the West as the SS6. Variations of the R7 rocket motor are still used as boosters for various manned and unmanned Soviet space flights. Korolyov was elected to the U.S.S.R. Academy of Sciences in 1958. His remains are buried in the Kremlin Wall.

Alexei Archipovich Leonov

Cosmonaut, born May 30, 1934. An air force lieutenant colonel, Leonov went to aviation school in 1953 and graduated in 1957, when he became a fighter pilot, a member of the Communist Party, and a parachutist. He was selected to be a cosmonaut in 1959. Leonov's only flight into space was on *Voskhod 2,* during which he became the first human to walk in space. During Leonov's 10-minute walk he was tethered to the spacecraft.

James Arthur Lovell, Jr.

U.S. astronaut, born March 25, 1928. A navy captain, Lovell was a 1952 graduate of the U.S. Naval Academy. He then became a test pilot, and in 1963, when NASA recruited him, Lovell was a flight instructor and safety officer. His first flight into orbit was aboard *Gemini 7,* which together with *Gemini 6* participated in the first rendezvous of two spacecraft in orbit. Lovell flew again on *Gemini 12,* the last flight of the Gemini series, and on *Apollo 8,* which earned him the distinction of becoming one of the first three humans to leave Earth for another celestial body—the Moon. *Apollo 8* was humankind's first manned flight to another celestial body. It completed ten lunar orbits on Christmas Day, 1968, thus paving the way for a lunar landing the following year. Lovell's final flight was the near-catastrophic *Apollo 13,* which he was commanding to the Moon when an explosion in an oxygen tank tore apart the spacecraft's service module 330,000 kilometers (205,000 miles) from Earth. The ingenuity of Lovell and his crew and the ground crew's support saved the astronauts and brought them back alive to Earth. Lovell resigned from the space program after *Apollo 13.*

Sharon Christa McAuliffe

U.S. astronaut, born September 2, 1948, died January 28, 1986. A civilian from Concorde, New Hampshire, chosen for a flight into space on the space shuttle *Challenger*, mission STS 51L. McAuliffe was 37 when, out of 11,000 applicants, she was selected by NASA to participate in the Teacher in Space program. She had been a schoolteacher for 15 years, teaching English, U.S. history, and economics. She was chosen for her enthusiasm for the U.S. space program and her ability to effectively communicate her experiences to young people. She was married to a lawyer and had two children. McAuliffe was to teach a class from space and keep a journal of her flight that she hoped would convey the "ordinary person's perspective" of space flight. McAuliffe, however, never made it into space, for on January 28, 1986, the shuttle *Challenger* exploded 73 seconds after lift-off, killing all seven of its astronauts. It was the worst disaster yet in humankind's space exploration efforts.

Ronald E. McNair

U.S. astronaut, born October 21, 1950, died January 28, 1986. A civilian, McNair earned a bachelor's degree in physics from North Carolina A & T State College, then earned a doctorate in physics from the Massachusetts Institute of Technology. NASA recruited him as an astronaut in 1978 and sent him into space as a mission specialist on the shuttle *Challenger*, mission STS 41B. His next flight was to have also been on the *Challenger*, in 1986, but 73 seconds after lift-off the *Challenger* exploded, killing all seven of its crew. It was the worst disaster yet in humankind's space exploration efforts. Ronald McNair was survived by a wife and son.

Edgar Dean Mitchell

U.S. astronaut, born September 17, 1930. A navy captain, Mitchell earned a master's degree in industrial management from the Carnegie Institute of Technology, now Carnegie-Mellon University, in Pittsburgh. He then joined the navy. In 1961, he earned a master's degree in aeronautical engineering from the U.S. Naval Postgraduate School, and then in 1964, he earned a doctorate in aeronautics and astronautics from the Massachusetts Institute of Technology. NASA recruited Mitchell in 1966 and sent him to the Moon in 1971 as the lunar module pilot of *Apollo 14*. Mitchell thus became the sixth human to set foot on another celestial body. The following year, Mitchell resigned from the navy and the space program.

Ellison S. Onizuka

U.S. astronaut, born June 24, 1946, died January 28, 1986. An air force major, Onizuka was a Japanese-American raised in Hawaii who dreamed of going to the Moon one day. Onizuka had bachelor's and master's degrees in aerospace engineering from the University of Colorado. He was a flight test engineer at Edwards Air Force Base when NASA recruited him as an astronaut in 1978. Onizuka's first flight into space was in 1985 as a mission specialist on the shuttle *Discovery*, mission STS 51C. His next mission was to have been STS 51L on the shuttle *Challenger* in 1986, but 73 seconds after lift-off the *Challenger* exploded, killing all seven of its crew. It was the worst disaster yet in humankind's space exploration efforts. Onizuka was survived by a wife and two children.

Viktor Ivanovich Patsayev

Cosmonaut, born June 19, 1933, died June 29, 1971. A civilian, Patsayev was a graduate of Penze Industrial Institute. He was a radio researcher and design engineer at the Central Aerological Observatory when he was recruited into the cosmonaut detachment in 1967. His first flight into space was aboard *Soyuz 11*, which docked with the famous *Salyut 1* space station. *Soyuz 11* remained in space for 24 days, during which Patsayev and his two crew mates conducted meteorological and plant-growth experiments. On June 29, 1971, Patsayev and his crew mates were killed during their descent back to Earth. It is believed that a valve opened accidentally during re-entry and depressurized the spacecraft's cabin. The cosmonauts were found dead upon recovery of the craft.

Judith A. Resnik

U.S. astronaut, born April 5, 1949, died January 28, 1986. A civilian, Resnik had a bachelor's degree in electrical engineering from Carnegie-Mellon University and a doctorate in electrical engineering from the University of Maryland. NASA recruited her as an astronaut in 1978 to be a mission specialist on shuttle missions. She became the second U.S. woman in space in 1984, when she flew into orbit for the first time on the shuttle *Discovery*, mission STS 41D. Her second flight into space would have been in 1986 on the shuttle *Challenger,* mission STS 51L, but the *Challenger* exploded 73 seconds after lift-off, killing all seven of its crew. It was the worst disaster yet in humankind's space exploration efforts. Resnik was divorced and had no children.

Sally Kristen Ride

U.S. astronaut, born May 26, 1951. Ride was the first U.S. woman in space. A civilian, Ride holds a bachelor of science degree in physics, a bachelor of arts in English, and both a master of science and a doctorate in X-ray physics from Stanford University. In 1978, Ride applied to NASA and was selected as one of the six women chosen to be mission specialists on upcoming shuttle missions. During this recruitment period NASA chose a total of 35 future astronauts. One of them was Dr. Steven Hawley, whom Ride married in 1982. Ride flew twice on the shuttle *Challenger:* first on mission STS 7, June 18–24, 1983, which made her the first U.S. woman in space, and again on mission STS 41G, August 5–13, 1984. On mission STS 7, Ride helped launch satellites and test the shuttle's remote manipulator arm. After the *Challenger* disaster of January 28, 1986, Ride was appointed to the president's commission investigating the accident. Shortly after the commission's work was over, Ride resigned from the astronaut program and returned to Stanford University.

Walter Marty Schirra, Jr.

U.S. astronaut, born March 12, 1923. Schirra was one of the first seven U.S. astronauts. A navy captain, he earned a bachelor of science degree from the U.S. Naval Academy in 1945. He flew combat missions during the Korean War and shot down one MiG. He participated in the development of the Sidewinder air-to-air missile and was then assigned to the carrier USS *Lexington,* then in the Pacific, as an operations officer. From that duty he went on to become a test pilot at Patuxent, Maryland. By the time he was selected as an astronaut, he had logged over 3,000 hours of flying time, over 1,700 of them in jets. Schirra piloted *Mercury-Sigma 7* into Earth orbit in 1962. In 1965, he commanded *Gemini 6* into orbit, and in 1968, he commanded *Apollo 7* into orbit, the flight during which he directed activities in space that paved the way for a lunar landing the following year.

Harrison Hagan Schmitt

U.S. astronaut, born July 3, 1935. A civilian, Schmitt earned a bachelor's degree in geology from the California Institute of Technology in 1957 and a doctorate in geology from Harvard University in 1964. He had worked for the U.S. Geological Survey in New Mexico and Montana before NASA recruited him as an astronaut in 1965. Schmitt's only mission into space was as the lunar module pilot on the historic *Apollo 17* in 1972, the seventh and as yet last manned lunar landing mission. Schmitt thus became the twelfth human to set foot on another celestial body.

Francis R. Scobee

U.S. astronaut, born May 19, 1939, died January 28, 1986. An air force major, Scobee had a bachelor's degree in aerospace engineering from the University of Arizona. During his 22-year service with the air force, Scobee flew combat missions in Vietnam and then became a test pilot. NASA recruited him into astronaut group eight in 1978. Scobee retired from the air force in 1980 and devoted full time to training as an astronaut. His first flight into space was in 1984 as a pilot of the shuttle *Challenger,* mission STS 41C. His second flight was to have been in 1986 as the commander of mission STS 51L on the shuttle *Challenger,* but 73 seconds after lift-off the *Challenger* exploded, killing all seven of its crew. It was the worst disaster yet in humankind's space exploration efforts. Scobee was survived by a wife and two children.

Alan Bartlett Shepard, Jr.

U.S. astronaut, born November 18, 1923. A navy rear admiral, Shepard was one of the first seven U.S. astronauts. Shepard earned a bachelor of science degree from the U.S. Naval Academy in 1944 and served briefly on a cruiser during the final days of World War II. He then opted for flight training and flew carrier missions in the Mediterranean, after which he went into test flying and participated in the development of the navy's in-flight refueling systems. He was also one of the test pilots who helped the navy assess landings and takeoffs from angled carrier decks. He tested the F3H Demon, the F8U Crusader, and the FL1F Tigercat jets. At the time he was selected as an astronaut, he was on the Atlantic Fleet commander-in-chief's staff and had logged over 3,600 hours of flying time. Shepard holds the distinction of being the first U.S. astronaut in space, a distinction he earned in 1961 when he piloted *Mercury 3* into a suborbital flight. In 1971, Shepard commanded humankind's fourth manned mission to the Moon, *Apollo 14*, and became the fifth human in history, and one of only 12 men so far, to walk on the Moon. (He is also known as the first human to have played golf on another celestial body.) On October 1, 1978, President Jimmy Carter honored Shepard with the Congressional Space Medal of Honor for his contribution to space exploration.

Donald Kent Slayton

U.S. astronaut, born March 1, 1924. A major in the air force reserve, Slayton was a pilot of medium bombers, such as the B25 and the A26, during World War II. He flew combat missions over both Germany and Japan. For a brief period after the war he was a flight instructor, but then he left the service to attend college at the University of Minnesota, where he earned a bachelor of science degree in aeronautical engineering in 1949. In 1951, he returned to the air force as a test pilot, then went to Germany as a fighter pilot. In 1955, he returned to the United States to train as a test pilot at Edwards Air Force Base, where he then tested virtually every new jet fighter for the air force. By the time he was selected as an astronaut, Slayton had logged over 3,400 hours of flying time, over 2,000 of them in jets. A minor heart irregularity, however, grounded Slayton until 1975, when he co-piloted the U.S. side of the historic Apollo-Soyuz link-up in space. This first international mission produced the first international handshake in space, between a U.S. astronaut and a U.S.S.R. cosmonaut.

Michael J. Smith

U.S. astronaut, born April 30, 1945, died January 28, 1986. A navy commander, Smith was a graduate of the U.S. Naval Academy with a bachelor's degree in naval sciences. He also held a master's degree in aeronautical engineering from the U.S. Naval Postgraduate School. Smith had flown combat missions in Vietnam and was an instructor at the Patuxent River Test Pilot School. NASA recruited him as an astronaut in 1980. His first flight into space would have been on the shuttle *Challenger,* mission STS 51L, in 1986, but 73 seconds after lift-off the *Challenger* exploded, killing all seven of its crew. It was the worst disaster yet in humankind's space exploration efforts. Smith was survived by a wife and three children.

Valentina Vladimirovna Tereshkova

Cosmonaut, born March 6, 1937. Tereshkova was the first woman in space. She was one of three children born into a peasant family, and her father died in World War II. After leaving school, Tereshkova worked in a textile factory, but she continued to attend night school as well as completing a correspondence course from a technical school. At the textile factory, she was elected to the factory committee, and later, she was also elected to the Communist Party. She was also a parachutist, with about 126 jumps to her credit by the time she applied for cosmonaut training in March 1962, when she was selected as one of the four women from whom the first woman in space would be chosen. The final decision was announced shortly before her flight into space on June 16, 1963, a little over a year after she had begun cosmonaut training. Tereshkova piloted *Vostok 6* into Earth orbit.

Konstantin Eduardovich Tsiolkovsky

Soviet rocket visionary, born September 17, 1857, died September 19, 1935. Tsiolkovsky was a Soviet schoolteacher who since his teens had been interested in space travel. He educated himself in astronomy, mathematics, and physics and then developed a theory of rocket propulsion. In 1898, Tsiolkovsky wrote the first paper ever written on rocketry and space travel. The paper was titled "Research into Interplanetary Space by Means of Rocket Power." Tsiolkovsky submitted the paper to the magazine *Science Survey*, but his work was not published until 1903. In the paper, Tsiolkovsky concluded that out of the known fuels, the only fuels that would pack enough energy to hurl humans into space are liquid fuels, particularly liquid hydrogen and its oxidizer, liquid oxygen. Furthermore, Tsiolkovsky concluded that for a successful ascent into space, rockets would have to be multistaged. When the fuel of one stage was consumed, the stage would be discarded and the next stage would fire, thus progressively increasing the rocket's velocity until it escaped into space. Tsiolkovsky's conclusions were, as we know today, correct. The U.S. space shuttle, for example, indeed uses liquid hydrogen for fuel and liquid oxygen as the fuel's oxidizer, and most rockets that humankind has launched into space have been multistaged. To honor

Tsiolkovsky, the Soviets launched humankind's exploration of space with *Sputnik 1* on October 4, 1957, the year of the one-hundredth anniversary of Tsiolkovsky's birth.

Vladislav Nikolayevich Volkov

Cosmonaut, born November 23, 1935, died June 29, 1971. A graduate of the Moscow Aviation Institute, Volkov was one of Sergei Korolyov's spacecraft design bureau engineers and was inducted into the cosmonaut detachment in 1966. Volkov first flew into space as a flight engineer aboard *Soyuz 7* in 1969, helping to perform welding experiments in space during this mission. His next flight was in 1971 aboard *Soyuz 11*. This mission lasted 24 days, during which Volkov and his two crew mates docked with the famous *Salyut 1* space station and performed meteorological and plant-growth experiments. On June 29, 1971, Volkov was killed during *Soyuz 11*'s descent back to Earth. It is believed that a valve opened accidentally during re-entry and depressurized the cabin, killing the three members of the crew.

Wernher von Braun

German rocket engineer, later a naturalized U.S. citizen, born March 23, 1912, died June 16, 1977. Von Braun was the brain behind the rocket motors of the U.S. space program from its inception to the *Apollo* spacecraft that boosted humans to the Moon. Born into a prosperous family in Germany, von Braun was given ample facilities in his young days to be curious about his surroundings and to satisfy his curiosity with knowledge. In 1925, he read the book titled *The Rocket into Interplanetary Space*, written by Hermann Oberth, a German pioneer in rocketry. Von Braun was so fascinated by the book that he dedicated the rest of his life to developing increasingly powerful rocket motors. In 1932, von Braun graduated as a mechanical engineer from the Berlin Institute of Technology. He enrolled at Berlin University for further studies and at the same time was inducted into the rocket research group of the Ordnance Department of the German armed forces. Two years later, he received his doctorate from Berlin University for his thesis on rocket motors. Von Braun went on to develop the A-4 rocket, which was later renamed V-2 by Hitler's Propaganda Ministry. (V-2 stood for Vengeance Weapon 2.) With the V-2, von Braun made Germany the world leader in rocket technology. Toward the end of World War II, therefore, the Allied armies raced for not only Berlin but also Peenemünde, where von Braun and his group of engineers had done most of their work on rockets. Von Braun and some of his group surrendered to U.S. forces and were brought to the United States to continue their work on rocket motors. In 1952, von Braun moved to Huntsville, Alabama, where he became technical director of the army's ballistic missile program. Here, he developed the basic rocket motors that later put humankind into space. Von Braun became a U.S. citizen in 1955, and in 1958, when NASA was officially formed, he became director of NASA's Marshall Space Flight Center in Huntsville. He then developed the Saturn I, Saturn IB, and Saturn V rocket motors, the last of which took humans to the Moon. Von Braun resigned from NASA in 1972 to join private industry and establish the National Space Institute, which promotes public awareness and support of space exploration.

Edward Higgins White, II

U.S. astronaut, born November 14, 1930, died January 27, 1967. White was the first U.S. astronaut to perform a space walk. An air force lieutenant colonel, White graduated from West Point Military Academy in 1952 and joined the air force. He then served in a fighter squadron in Germany. In 1959, he earned a master of science degree in aeronautical engineering from the University of Michigan at Ann Arbor, and then he graduated from the Air Force Test Pilot School at Edwards Air Force Base. NASA recruited White in 1962 during its second round of astronaut selections. White first flew in space on *Gemini 4*, during which he became the first U.S. astronaut and the second person in history to perform a space walk. He also became the first person to use a hand-held maneuvering unit to propel himself in space. White's space walk lasted 21 minutes. White was chosen to fly again on *Apollo 1*, and he was training for that mission with Virgil Grissom and Roger Chaffee when an electrical short circuit ignited the oxygen-rich atmosphere of the spacecraft and killed all three astronauts. This was the first disaster of the U.S. space exploration program.

John Watts Young

U.S. astronaut, born September 24, 1930. A navy captain, Young graduated in 1952 with a bachelor's degree in aeronautical engineering from the Georgia Institute of Technology. He then joined the navy and served in Korea. In 1962, he set two altitude records in an F4B jet. NASA recruited Young that same year. Young's first flight into space was on *Gemini 3* in 1965, on which he was Virgil Grissom's copilot. For this flight, President Lyndon Johnson conferred on Young the Exceptional Service Medal. Young orbited Earth again in 1966 on *Gemini 10,* which docked with an Agena target vehicle. Young took to space again in 1969 on *Apollo 10,* the mission that was the final precursor to the first lunar landing. Young orbited the Moon while his two crew mates descended to within 1,500 meters (5,000 feet) of the lunar surface. Young went back to the Moon in 1972, this time as commander of *Apollo 16.* During that historic flight he became the ninth human to set foot on another celestial body. In 1976, Young retired from the navy, but he stayed with NASA and became chief of the office of astronauts. In 1981, Young launched into space again on yet another historic flight—the maiden voyage of the space shuttle. Young commanded the shuttle *Columbia* with pilot Robert Crippen on April 12, 1981. In 1983, Young commanded a shuttle flight once more, during which he placed *Spacelab,* a permanent space station, into orbit. For his achievements, leadership, and courage throughout a space career that started with Gemini and spanned through Apollo into the shuttle program, Young was awarded the Congressional Space Medal of Honor on May 19, 1981, by President Ronald Reagan.

4

Facts and Data

THIS CHAPTER PRESENTS SUMMARIES of space exploration activities in tabular form. The tables have been compiled from regularly published reports such as NASA's *Satellite Situation Report*, TRW *Space Log*, and other sources listed in Chapter 5. The tables are organized as follows:

Unmanned Missions, Earth Orbit

Table 1: Unmanned Soviet Missions that Achieved Earth Orbit

Table 2: Unmanned U.S. Missions that Achieved Earth Orbit

Table 3: Unmanned Missions of Other Countries that Achieved Earth Orbit

Table 4: Unmanned International Missions that Achieved Earth Orbit

Manned Missions, Earth Orbit

Table 5: Manned Soviet Missions that Achieved Earth Orbit

Table 6: Manned U.S. Missions that Achieved Earth Orbit

Lunar Missions

Table 7: Unmanned Lunar Missions

Table 8: Manned Lunar Missions

Other

Table 9: Space Explorers, Their Missions, and Mission Durations

Table 10: Interplanetary Missions

Table 11: Winners of U.S. Congressional Space Medal of Honor

Table 12: Missions that Failed to Achieve Earth Orbit

Table 13: Space Exploration Fatalities

The organization of the tables gives credit to the U.S.S.R. for launching space exploration with the first unmanned mission that achieved Earth orbit. Soviet unmanned missions, therefore, are listed first. The U.S.S.R. is also given credit for putting the first human into orbit. For manned missions also, therefore, Soviet missions are listed first. The tables provide information up to June 1988 for Soviet missions and January 1990 for U.S. missions. Satellite decay information, regardless of country, is current up to January 1990.

TABLE 1   Unmanned Soviet Missions that Achieved Earth Orbit

MISSION	LAUNCH	LANDING, RECOVERY, OR RE-ENTRY	NOTES
ASAT-KOSMOS: Anti-satellite Kosmos			
Kosmos 185	27 Oct 1967	14 Jan 1969	
Kosmos 217	24 Apr 1968	26 Apr 1968	
Kosmos 248	19 Oct 1968	26 Feb 1980	Target *
Kosmos 249	20 Oct 1968		
Kosmos 252	1 Nov 1968		
Kosmos 291	6 Aug 1969	8 Sep 1969	
Kosmos 373	20 Oct 1970	8 Mar 1980	Target
Kosmos 374	23 Oct 1970		
Kosmos 375	30 Oct 1970		
Kosmos 394	9 Feb 1971		Target
Kosmos 397	25 Feb 1971		
Kosmos 400	18 Mar 1971		Target
Kosmos 404	4 Apr 1971	4 Apr 1971	
Kosmos 459	29 Nov 1971	27 Dec 1971	Target
Kosmos 462	3 Dec 1971	4 Apr 1975	
Kosmos 521	29 Sep 1972		Target
Kosmos 803	12 Feb 1976		Target
Kosmos 804	16 Feb 1976	16 Feb 1976	
Kosmos 814	13 Apr 1976	13 Apr 1976	
Kosmos 839	8 Jul 1976		Target
Kosmos 843	21 Jul 1976	21 Jul 1976	
Kosmos 880	9 Dec 1976	8 Oct 1979	Target
Kosmos 886	27 Dec 1976		
Kosmos 909	19 May 1977		Target
Kosmos 910	23 May 1977	23 May 1977	
Kosmos 918	18 Jun 1977	18 Jun 1977	
Kosmos 959	21 Oct 1977	30 Nov 1977	Target
Kosmos 961	26 Oct 1977	26 Oct 1977	
Kosmos 967	13 Dec 1977		Target
Kosmos 970	21 Dec 1977		
Kosmos 1009	19 May 1978	19 May 1978	
Kosmos 1171	3 Apr 1980		Target
Kosmos 1174	18 Apr 1980		
Kosmos 1241	21 Jan 1981		Target
Kosmos 1243	2 Feb 1981	2 Feb 1981	
Kosmos 1258	14 Mar 1981	14 Mar 1981	
Kosmos 1375	6 Jun 1982		Target
Kosmos 1379	18 Jun 1982	18 Jun 1982	
Kosmos 1427	29 Dec 1982		
Kosmos 1450	6 Apr 1983		
Kosmos 1502	5 Oct 1983	29 Aug 1985	

* **"Target" denotes a robotic vehicle used for experimental purposes by another vehicle.**

MISSION	LAUNCH	LANDING, RECOVERY, OR RE-ENTRY	NOTES
EKRAN (Communications)			
Ekran 1	26 Oct 1976		
Ekran 2	20 Sep 1977		
Ekran 3	21 Feb 1979		
Ekran 4	3 Oct 1979		
Ekran 5	14 Jul 1980		
Ekran 6	26 Dec 1980		
Ekran 7	26 Jun 1981		
Ekran 8	5 Feb 1982		
Ekran 9	16 Sep 1982		
Ekran 10	12 Mar 1983		
Ekran 11	29 Sep 1983		
Ekran 12	16 Mar 1984		
Ekran 13	24 Aug 1984		
Ekran 14	22 Mar 1985		
Ekran 15	24 May 1986		
Ekran 16	3 Sep 1987		
Ekran 17	27 Dec 1987		
ELEKTRON (Science, Technology)			
Elektron 1	30 Jan 1964		
Elektron 2	30 Jan 1964		
Elektron 3	11 Jul 1964		
Elektron 4	11 Jul 1964	12 Oct 1983	
ELINT-KOSMOS: Electronic Intelligence Kosmos			
Kosmos 103	28 Dec 1965		
Kosmos 151	24 Mar 1967		
Kosmos 189	30 Oct 1967	8 Jun 1978	
Kosmos 200	20 Jan 1968	24 Feb 1973	
Kosmos 236	27 Aug 1968		
Kosmos 250	31 Oct 1968	15 Feb 1978	
Kosmos 269	5 Mar 1969	21 Oct 1978	
Kosmos 315	20 Dec 1969	25 Mar 1979	
Kosmos 330	7 Apr 1970	12 Jun 1979	
Kosmos 358	20 Aug 1970		
Kosmos 387	16 Dec 1970	19 Jan 1980	
Kosmos 389	18 Dec 1970		
Kosmos 395	17 Feb 1971	6 Apr 1980	
Kosmos 405	7 Apr 1971		
Kosmos 425	29 May 1971	15 Jan 1980	
Kosmos 436	7 Sep 1971	4 Jan 1980	
Kosmos 437	10 Sep 1971	29 Mar 1980	
Kosmos 460	30 Nov 1971	5 Mar 1980	
Kosmos 476	1 Mar 1972		
Kosmos 479	22 Mar 1972	13 Apr 1980	
Kosmos 500	10 Jul 1972	29 Mar 1980	
Kosmos 536	3 Nov 1972	20 Jul 1980	

TABLE 1 49

MISSION	LAUNCH	LANDING, RECOVERY, OR RE-ENTRY	NOTES
Kosmos 542	28 Dec 1972	9 Oct 1983	
Kosmos 544	20 Jan 1973	15 Jun 1980	
Kosmos 549	28 Feb 1973	29 Jun 1980	
Kosmos 582	28 Aug 1973	5 Sep 1980	
Kosmos 604	29 Oct 1973		
Kosmos 610	27 Nov 1973	15 Sep 1980	
Kosmos 631	6 Feb 1974	3 Oct 1980	
Kosmos 655	21 May 1974	19 Nov 1980	
Kosmos 661	21 Jun 1974	27 Aug 1980	
Kosmos 673	16 Aug 1974		
Kosmos 698	18 Dec 1974	9 Dec 1980	
Kosmos 707	5 Feb 1975	7 Sep 1980	
Kosmos 744	20 Jun 1975		
Kosmos 749	4 Jul 1975	26 Sep 1980	
Kosmos 756	22 Aug 1975		
Kosmos 781	21 Nov 1975	26 Nov 1980	
Kosmos 787	6 Jan 1976	12 Dec 1980	
Kosmos 790	22 Jan 1976	12 Nov 1980	
Kosmos 808	16 Mar 1976		
Kosmos 812	6 Apr 1976	30 Oct 1980	
Kosmos 845	27 Jul 1976	15 Nov 1980	
Kosmos 851	27 Aug 1976	5 Aug 1989	
Kosmos 870	2 Dec 1976	20 Dec 1980	
Kosmos 895	26 Feb 1977		
Kosmos 899	24 Mar 1977	19 Oct 1980	
Kosmos 924	4 Jul 1977	10 Feb 1981	
Kosmos 925	7 Jul 1977		
Kosmos 955	20 Sep 1977		
Kosmos 960	25 Oct 1977	22 Oct 1980	
Kosmos 975	10 Jan 1978		
Kosmos 1005	12 May 1978		
Kosmos 1008	17 May 1978	3 Jan 1981	
Kosmos 1043	10 Oct 1978		
Kosmos 1062	15 Dec 1978	20 Apr 1981	
Kosmos 1063	19 Dec 1978		
Kosmos 1077	13 Feb 1979		
Kosmos 1093	14 Apr 1979		
Kosmos 1114	11 Jul 1979	26 Dec 1981	
Kosmos 1116	20 Jul 1979		
Kosmos 1143	26 Oct 1979		
Kosmos 1145	27 Nov 1979		
Kosmos 1154	30 Jan 1980		
Kosmos 1184	4 Jun 1980		
Kosmos 1206	15 Aug 1980		
Kosmos 1222	21 Nov 1980		
Kosmos 1242	27 Jan 1981		
Kosmos 1271	19 May 1981		
Kosmos 1315	13 Oct 1981		

MISSION	LAUNCH	LANDING, RECOVERY, OR RE-ENTRY	NOTES
Kosmos 1340	19 Feb 1982		
Kosmos 1345	31 Mar 1982	27 Sep 1989	
Kosmos 1346	31 Mar 1982		
Kosmos 1351	21 Apr 1982	14 Mar 1983	
Kosmos 1356	5 May 1982		
Kosmos 1397	29 Jul 1982	18 May 1983	
Kosmos 1400	5 Aug 1982		
Kosmos 1418	21 Oct 1982	30 Sep 1983	
Kosmos 1437	20 Jan 1983		
Kosmos 1441	16 Feb 1983		
Kosmos 1453	19 Apr 1983		
Kosmos 1455	23 Apr 1983		
Kosmos 1465	26 May 1983	23 Jan 1985	
Kosmos 1470	23 Jun 1983		
Kosmos 1494	31 Aug 1983	26 Sep 1985	
Kosmos 1501	30 Sep 1983		
Kosmos 1515	15 Dec 1983		
Kosmos 1534	26 Jan 1984		
Kosmos 1536	8 Feb 1984		
Kosmos 1544	15 Mar 1984		
Kosmos 1606	18 Oct 1984		
Kosmos 1626	24 Jan 1985		
Kosmos 1633	5 Mar 1985		
Kosmos 1656	30 May 1985		
Kosmos 1674	8 Aug 1985		
Kosmos 1697	22 Oct 1985		
Kosmos 1703	22 Nov 1985		
Kosmos 1707	12 Dec 1985		
Kosmos 1726	17 Jan 1986		
Kosmos 1733	19 Feb 1986		
Kosmos 1743	15 May 1986		
Kosmos 1758	12 Jun 1986		
Kosmos 1782	30 Sep 1986		
Kosmos 1805	10 Dec 1986		
Kosmos 1812	14 Jan 1987		
Kosmos 1825	3 Mar 1987		
Kosmos 1833	18 Mar 1987		
Kosmos 1842	27 Apr 1987		
Kosmos 1844	13 May 1987		
Kosmos 1862	1 Jul 1987		
Kosmos 1892	20 Oct 1987		

FOBS-KOSMOS: Fractional Orbit Bombardment Systems Kosmos

Kosmos 139	25 Jan 1967	25 Jan 1967	
Kosmos 160	17 May 1967	18 May 1967	
Kosmos 169	17 Jul 1967	17 Jul 1967	
Kosmos 170	31 Jul 1967	31 Jul 1967	
Kosmos 171	8 Aug 1967	8 Aug 1967	
Kosmos 178	19 Sep 1967	19 Sep 1967	

TABLE 1   51

MISSION	LAUNCH	LANDING, RECOVERY, OR RE-ENTRY	NOTES
Kosmos 179	22 Sep 1967	22 Sep 1967	
Kosmos 183	18 Oct 1967	18 Oct 1967	
Kosmos 187	28 Oct 1967	28 Oct 1967	
Kosmos 218	25 Apr 1968	25 Apr 1968	
Kosmos 244	2 Oct 1968	2 Oct 1968	
Kosmos 298	15 Sep 1969	15 Sep 1969	
Kosmos 316	23 Dec 1969	28 Aug 1970	
Kosmos 354	28 Jul 1970	28 Jul 1970	
Kosmos 365	25 Sep 1970	25 Sep 1970	
Kosmos 433	8 Aug 1971	10 Aug 1971	

GORIZONT (Communications)

Gorizont 2	5 Jul 1979		
Gorizont 3	28 Dec 1979		
Gorizont 4	14 Jun 1980		
Gorizont 5	15 Mar 1982		
Gorizont 6	20 Oct 1982		
Gorizont 7	1 Jul 1983		
Gorizont 8	30 Nov 1983		
Gorizont 9	22 Apr 1984		
Gorizont 10	1 Aug 1984		
Gorizont 11	18 Jan 1985		
Gorizont 12	10 Jun 1986		
Gorizont 13	18 Nov 1986		
Gorizont 14	11 May 1987		

ISKRA (Communications)

Iskra 1	10 Jul 1981	7 Oct 1981	
Iskra 2	17 May 1982	9 Jul 1982	
Iskra 3	18 Nov 1982	16 Dec 1982	

KOSMOS (Communications)

Kosmos 38	18 Aug 1964	8 Nov 1964	
Kosmos 39	18 Aug 1964	17 Nov 1964	
Kosmos 40	18 Aug 1964	18 Nov 1964	
Kosmos 42	22 Aug 1964	19 Dec 1965	
Kosmos 54	21 Feb 1965	15 Sep 1968	
Kosmos 55	21 Feb 1965	2 Feb 1968	
Kosmos 56	21 Feb 1965	2 Nov 1967	
Kosmos 61	15 Mar 1965	15 Jan 1968	
Kosmos 62	15 Mar 1965	24 Sep 1968	
Kosmos 63	15 Mar 1965	4 Nov 1967	
Kosmos 71	16 Jul 1965	11 Aug 1970	
Kosmos 72	16 Jul 1965	24 Aug 1979	
Kosmos 73	16 Jul 1965	20 Mar 1974	
Kosmos 74	16 Jul 1965	13 Dec 1979	
Kosmos 75	16 Jul 1965	28 Sep 1979	
Kosmos 80	3 Sep 1975		
Kosmos 81	3 Sep 1975		
Kosmos 82	3 Sep 1975		

MISSION	LAUNCH	LANDING, RECOVERY, OR RE-ENTRY	NOTES
Kosmos 83	3 Sep 1975		
Kosmos 84	3 Sep 1975		
Kosmos 86	18 Sep 1975		
Kosmos 87	18 Sep 1975		
Kosmos 88	18 Sep 1975		
Kosmos 89	18 Sep 1975		
Kosmos 90	18 Sep 1975		
Kosmos 1366	18 May 1982		
Kosmos 1540	2 Mar 1984		
Kosmos 1546	29 Mar 1984		
Kosmos 1629	21 Feb 1985		
Kosmos 1700	25 Oct 1985		
Kosmos 1738	4 Apr 1986		
Kosmos 1888	1 Oct 1987		
Kosmos 1894	28 Oct 1987		
Kosmos 1897	25 Nov 1987		

KOSMOS (Early Warning Satellites)

MISSION	LAUNCH	LANDING, RECOVERY, OR RE-ENTRY	NOTES
Kosmos 174	31 Aug 1967	30 Dec 1968	
Kosmos 260	16 Dec 1968	9 Jul 1973	
Kosmos 520	19 Sep 1972		
Kosmos 606	2 Nov 1973		
Kosmos 665	29 Jun 1974		
Kosmos 706	30 Jan 1975		
Kosmos 775	8 Oct 1975		
Kosmos 862	22 Oct 1976		
Kosmos 903	11 Apr 1977		
Kosmos 917	16 Jun 1977		
Kosmos 931	20 Jul 1977		
Kosmos 1024	28 Jun 1978		
Kosmos 1030	6 Sep 1978		
Kosmos 1109	27 Jun 1979		
Kosmos 1124	28 Aug 1979		
Kosmos 1172	12 Apr 1980		
Kosmos 1188	14 Jun 1980		
Kosmos 1191	2 Jul 1980		
Kosmos 1217	24 Oct 1980		
Kosmos 1223	27 Nov 1980		
Kosmos 1247	19 Feb 1981		
Kosmos 1261	31 Mar 1981		
Kosmos 1278	19 Jun 1981		
Kosmos 1285	4 Aug 1981		
Kosmos 1317	31 Oct 1981		
Kosmos 1341	3 Mar 1982		
Kosmos 1348	7 Apr 1982		
Kosmos 1367	20 May 1982		
Kosmos 1382	25 Jun 1982		
Kosmos 1409	22 Sep 1982		
Kosmos 1456	25 Apr 1983		

TABLE 1   53

MISSION	LAUNCH	LANDING, RECOVERY, OR RE-ENTRY	NOTES
Kosmos 1481	8 Jul 1983		
Kosmos 1518	28 Dec 1983	8 Feb 1984	
Kosmos 1541	6 Mar 1984		
Kosmos 1547	4 Apr 1984		
Kosmos 1569	6 Jun 1984		
Kosmos 1581	3 Jul 1984		
Kosmos 1586	2 Aug 1984		
Kosmos 1596	7 Sep 1984		
Kosmos 1604	4 Oct 1984		
Kosmos 1658	11 Jun 1985		
Kosmos 1661	18 Jun 1985		
Kosmos 1675	12 Aug 1985		
Kosmos 1684	24 Sep 1985		
Kosmos 1687	30 Sep 1985		
Kosmos 1698	22 Oct 1985		
Kosmos 1701	9 Nov 1985		
Kosmos 1729	1 Feb 1986		
Kosmos 1761	5 Jul 1986		
Kosmos 1774	28 Aug 1986		
Kosmos 1785	15 Oct 1986		
Kosmos 1793	20 Nov 1986		
Kosmos 1806	12 Dec 1986		
Kosmos 1849	4 Jun 1987		
Kosmos 1851	11 Jun 1987		
Kosmos 1903	21 Dec 1987		

KOSMOS (Failed Attempts of Various Missions)

MISSION	LAUNCH	LANDING, RECOVERY, OR RE-ENTRY	NOTES
Kosmos 21	11 Nov 1963	14 Nov 1963	Venera 2 first attempt
Kosmos 27	27 Mar 1964	28 Mar 1964	Venera 2 second attempt
Kosmos 60	12 Mar 1965	17 Mar 1965	Luna 5 first attempt
Kosmos 96	23 Nov 1965	9 Dec 1965	Venera 4 first attempt
Kosmos 111	1 Mar 1966	3 Mar 1966	Luna 10 first attempt
Kosmos 146	10 Mar 1967	18 Mar 1967	Zond 4 first attempt
Kosmos 154	8 Apr 1967	10 Apr 1967	Zond 4 second attempt
Kosmos 159	17 May 1967	11 Nov 1967	Zond 4 third attempt
Kosmos 167	17 Jun 1967	25 Jun 1967	Venera 5 first attempt
Kosmos 300	23 Sep 1969	27 Sep 1969	Luna 16 first attempt
Kosmos 305	22 Oct 1969	24 Oct 1969	Luna 16 second attempt
Kosmos 359	22 Aug 1970	6 Nov 1970	Venera 8 first attempt
Kosmos 419	10 May 1971	12 May 1971	Mars 2 first attempt
Kosmos 482	31 Mar 1972	5 May 1981	Venera 9 first attempt
Kosmos 557	11 May 1973	22 May 1973	Salyut 3 first attempt
Kosmos 837	1 Jul 1976	18 Nov 1981	Molniya 2-16 first attempt
Kosmos 853	1 Sep 1976	31 Dec 1976	Molniya 2-16 second attempt
Kosmos 1072	16 Jan 1979		Kosmos 1089 first attempt
Kosmos 1164	12 Feb 1980	12 Jan 1981	Molniya 3-13 first attempt
Kosmos 1175	18 Apr 1980	28 May 1980	Molniya 3-13 second attempt

MISSION	LAUNCH	LANDING, RECOVERY, OR RE-ENTRY	NOTES
Kosmos 1305	11 Sep 1981		Molniya 3-17 first attempt
Kosmos 1380	18 Jun 1982	27 Jun 1982	Kosmos 1383 first attempt
Kosmos 1423	8 Dec 1982		Molniya 1-56 first attempt
Kosmos 1626	27 Nov 1984	31 Jan 1985	Kosmos 1626 first attempt
Kosmos 1625	23 Jan 1985	25 Jan 1985	Kosmos 1646 first attempt
Kosmos 1714	28 Dec 1985	27 Feb 1986	Kosmos 1726 first attempt
Kosmos 1783	3 Oct 1986		Kosmos 1785 first attempt
Kosmos 1817	29 Jan 1987	18 Mar 1987	Ekran 16 first attempt
Kosmos 1838	24 Apr 1987		Kosmos 1883 first attempt
Kosmos 1839	24 Apr 1987		Kosmos 1883 second attempt
Kosmos 1840	24 Apr 1987		Kosmos 1883 third attempt

KOSMOS (Reconnaissance, Surveillance)

MISSION	LAUNCH	LANDING, RECOVERY, OR RE-ENTRY
Kosmos 4	26 Apr 1962	29 Apr 1962
Kosmos 7	28 Jul 1962	1 Aug 1962
Kosmos 9	27 Sep 1962	1 Oct 1962
Kosmos 10	17 Oct 1962	21 Oct 1962
Kosmos 12	22 Dec 1962	30 Dec 1962
Kosmos 13	21 Mar 1963	29 Mar 1963
Kosmos 15	22 Apr 1963	27 Apr 1963
Kosmos 16	28 Apr 1963	8 May 1963
Kosmos 18	24 May 1963	2 Jun 1963
Kosmos 20	18 Oct 1963	28 Oct 1963
Kosmos 22	16 Nov 1963	22 Nov 1963
Kosmos 24	19 Dec 1963	28 Dec 1963
Kosmos 28	4 Apr 1964	12 Apr 1964
Kosmos 29	25 Apr 1964	3 May 1964
Kosmos 30	18 May 1964	26 May 1964
Kosmos 32	10 Jun 1964	18 Jun 1964
Kosmos 33	23 Jun 1964	1 Jul 1964
Kosmos 34	1 Jul 1964	9 Jul 1964
Kosmos 35	15 Jul 1964	23 Jul 1964
Kosmos 37	14 Aug 1964	22 Aug 1964
Kosmos 45	13 Sep 1964	18 Sep 1964
Kosmos 46	24 Sep 1964	2 Oct 1964
Kosmos 48	14 Oct 1964	20 Oct 1964
Kosmos 50	28 Oct 1964	5 Nov 1964
Kosmos 52	11 Jan 1965	19 Jan 1965
Kosmos 59	7 Mar 1965	15 Mar 1965
Kosmos 64	25 Mar 1965	2 Apr 1965
Kosmos 65	17 Apr 1965	25 Apr 1965
Kosmos 66	7 May 1965	15 May 1965
Kosmos 67	25 May 1965	2 Jun 1965
Kosmos 68	15 Jun 1965	23 Jun 1965
Kosmos 69	25 Jun 1965	3 Jul 1965
Kosmos 77	3 Aug 1965	11 Aug 1965
Kosmos 78	14 Aug 1965	22 Aug 1965
Kosmos 79	25 Aug 1965	2 Sep 1965

TABLE 1   55

MISSION	LAUNCH	LANDING, RECOVERY, OR RE-ENTRY	NOTES
Kosmos 85	9 Sep 1965	17 Sep 1965	
Kosmos 91	23 Sep 1965	1 Oct 1965	
Kosmos 92	16 Oct 1965	24 Oct 1965	
Kosmos 94	28 Oct 1965	5 Nov 1965	
Kosmos 98	27 Nov 1965	5 Dec 1965	
Kosmos 99	10 Dec 1965	18 Dec 1965	
Kosmos 104	7 Jan 1966	15 Jan 1966	
Kosmos 105	22 Jan 1966	30 Jan 1966	
Kosmos 107	10 Feb 1966	18 Feb 1966	
Kosmos 109	19 Feb 1966	27 Feb 1966	
Kosmos 112	17 Mar 1966	25 Mar 1966	
Kosmos 113	21 Mar 1966	29 Mar 1966	
Kosmos 114	6 Apr 1966	14 Apr 1966	
Kosmos 115	20 Apr 1966	28 Apr 1966	
Kosmos 117	6 May 1966	14 May 1966	
Kosmos 120	8 Jun 1966	16 Jun 1966	
Kosmos 121	17 Jun 1966	25 Jun 1966	
Kosmos 124	14 Jul 1966	22 Jul 1966	
Kosmos 126	28 Jul 1966	6 Aug 1966	
Kosmos 127	8 Aug 1966	16 Aug 1966	
Kosmos 128	27 Aug 1966	4 Sep 1966	
Kosmos 129	14 Oct 1966	21 Oct 1966	
Kosmos 130	20 Oct 1966	28 Oct 1966	
Kosmos 131	12 Nov 1966	20 Nov 1966	
Kosmos 132	19 Nov 1966	27 Nov 1966	
Kosmos 134	3 Dec 1966	11 Dec 1966	
Kosmos 136	19 Dec 1966	27 Dec 1966	
Kosmos 138	19 Jan 1967	27 Jan 1967	
Kosmos 141	8 Feb 1967	16 Feb 1967	
Kosmos 143	27 Feb 1967	7 Mar 1967	
Kosmos 147	13 Mar 1967	21 Mar 1967	
Kosmos 150	22 Mar 1967	30 Mar 1967	
Kosmos 153	4 Apr 1967	12 Apr 1967	
Kosmos 155	12 Apr 1967	20 Apr 1967	
Kosmos 157	12 May 1967	20 May 1967	
Kosmos 161	22 May 1967	30 May 1967	
Kosmos 162	1 Jun 1967	9 Jun 1967	
Kosmos 164	8 Jun 1967	14 Jun 1967	
Kosmos 168	4 Jul 1967	12 Jul 1967	
Kosmos 172	9 Aug 1967	17 Aug 1967	
Kosmos 175	11 Sep 1967	19 Sep 1967	
Kosmos 177	16 Sep 1967	24 Sep 1967	
Kosmos 180	26 Sep 1967	4 Oct 1967	
Kosmos 181	11 Oct 1967	18 Oct 1967	
Kosmos 182	16 Oct 1967	24 Oct 1967	
Kosmos 190	3 Nov 1967	11 Nov 1967	
Kosmos 193	25 Nov 1967	3 Dec 1967	
Kosmos 194	3 Dec 1967	11 Dec 1967	

MISSION	LAUNCH	LANDING, RECOVERY, OR RE-ENTRY	NOTES
Kosmos 195	16 Dec 1967	23 Dec 1967	
Kosmos 199	16 Jan 1968	1 Feb 1968	
Kosmos 201	6 Feb 1968	14 Feb 1968	
Kosmos 205	5 Mar 1968	13 Mar 1968	
Kosmos 207	16 Mar 1968	24 Mar 1968	
Kosmos 208	21 Mar 1968	2 Apr 1968	
Kosmos 210	3 Apr 1968	11 Apr 1968	
Kosmos 214	18 Apr 1968	26 Apr 1968	
Kosmos 216	20 Apr 1968	28 Apr 1968	
Kosmos 223	1 Jun 1968	9 Jun 1968	
Kosmos 224	4 Jun 1968	12 Jun 1968	
Kosmos 227	18 Jun 1968	26 Jun 1968	
Kosmos 228	21 Jun 1968	3 Jul 1968	
Kosmos 229	26 Jun 1968	4 Jul 1968	
Kosmos 231	10 Jul 1968	18 Jul 1968	
Kosmos 232	16 Jul 1968	24 Jul 1968	
Kosmos 234	30 Jul 1968	5 Aug 1968	
Kosmos 235	9 Aug 1968	17 Aug 1968	
Kosmos 237	27 Aug 1968	4 Sep 1968	
Kosmos 239	5 Sep 1968	13 Sep 1968	
Kosmos 240	14 Sep 1968	21 Sep 1968	
Kosmos 241	16 Sep 1968	24 Sep 1968	
Kosmos 243	23 Sep 1968	4 Oct 1968	
Kosmos 246	7 Oct 1968	12 Oct 1968	
Kosmos 247	11 Oct 1968	19 Oct 1968	
Kosmos 251	31 Oct 1968	18 Nov 1968	
Kosmos 253	13 Nov 1968	18 Nov 1968	
Kosmos 254	21 Nov 1968	29 Nov 1968	
Kosmos 255	29 Nov 1968	7 Dec 1968	
Kosmos 258	10 Dec 1968	18 Dec 1968	
Kosmos 263	12 Jan 1969	20 Jan 1969	
Kosmos 264	23 Jan 1969	5 Feb 1969	
Kosmos 266	25 Feb 1969	5 Mar 1969	
Kosmos 267	26 Feb 1969	6 Mar 1969	
Kosmos 270	6 Mar 1969	14 Mar 1969	
Kosmos 271	15 Mar 1969	21 Mar 1969	
Kosmos 273	22 Mar 1969	30 Mar 1969	
Kosmos 274	24 Mar 1969	1 Apr 1969	
Kosmos 276	4 Apr 1969	11 Apr 1969	
Kosmos 278	9 Apr 1969	17 Apr 1969	
Kosmos 279	15 Apr 1969	23 Apr 1969	
Kosmos 280	23 Apr 1969	6 May 1969	
Kosmos 281	13 May 1969	21 May 1969	
Kosmos 282	20 May 1969	28 May 1969	
Kosmos 284	29 May 1969	6 Jun 1969	
Kosmos 286	15 Jun 1969	23 Jun 1969	
Kosmos 287	24 Jun 1969	2 Jul 1969	
Kosmos 288	27 Jun 1969	5 Jul 1969	

TABLE 1   57

MISSION	LAUNCH	LANDING, RECOVERY, OR RE-ENTRY	NOTES
Kosmos 289	10 Jul 1969	15 Jul 1969	
Kosmos 290	22 Jul 1969	30 Jul 1969	
Kosmos 293	16 Aug 1969	28 Aug 1969	
Kosmos 294	19 Aug 1969	27 Aug 1969	
Kosmos 296	29 Aug 1969	6 Sep 1969	
Kosmos 297	2 Sep 1969	10 Sep 1969	
Kosmos 299	18 Sep 1969	22 Sep 1969	
Kosmos 301	24 Sep 1969	2 Oct 1969	
Kosmos 302	17 Oct 1969	25 Oct 1969	
Kosmos 306	24 Oct 1969	5 Nov 1969	
Kosmos 309	12 Nov 1969	20 Nov 1969	
Kosmos 310	15 Nov 1969	23 Nov 1969	
Kosmos 313	3 Dec 1969	15 Dec 1969	
Kosmos 317	23 Dec 1969	5 Jan 1970	
Kosmos 318	9 Jan 1970	21 Jan 1970	
Kosmos 322	21 Jan 1970	29 Jan 1970	
Kosmos 323	10 Feb 1970	18 Feb 1970	
Kosmos 325	4 Mar 1970	12 Mar 1970	
Kosmos 326	14 Mar 1970	21 Mar 1970	
Kosmos 328	27 Mar 1970	9 Apr 1970	
Kosmos 329	3 Apr 1970	15 Apr 1970	
Kosmos 331	8 Apr 1970	16 Apr 1970	
Kosmos 333	15 Apr 1970	28 Apr 1970	
Kosmos 344	12 May 1970	20 May 1970	
Kosmos 345	20 May 1970	28 May 1970	
Kosmos 346	10 Jun 1970	17 Jun 1970	
Kosmos 349	17 Jun 1970	25 Jun 1970	
Kosmos 350	26 Jun 1970	8 Jul 1970	
Kosmos 352	7 Jul 1970	15 Jul 1970	
Kosmos 353	9 Jul 1970	21 Jul 1970	
Kosmos 355	7 Aug 1970	15 Aug 1970	
Kosmos 360	29 Aug 1970	8 Sep 1970	
Kosmos 361	8 Sep 1970	21 Sep 1970	
Kosmos 363	17 Sep 1970	29 Sep 1970	
Kosmos 364	22 Sep 1970	2 Oct 1970	
Kosmos 366	1 Oct 1970	13 Oct 1970	
Kosmos 368	8 Oct 1970	14 Oct 1970	
Kosmos 370	9 Oct 1970	22 Oct 1970	
Kosmos 376	30 Oct 1970	12 Nov 1970	
Kosmos 377	11 Nov 1970	23 Nov 1970	
Kosmos 383	3 Dec 1970	16 Dec 1970	
Kosmos 384	10 Dec 1970	22 Dec 1970	
Kosmos 386	15 Dec 1970	28 Dec 1970	
Kosmos 390	12 Jan 1971	25 Jan 1971	
Kosmos 392	21 Jan 1971	2 Feb 1971	
Kosmos 396	18 Feb 1971	3 Mar 1971	
Kosmos 399	3 Mar 1971	17 Mar 1971	
Kosmos 401	27 Mar 1971	9 Apr 1971	

MISSION	LAUNCH	LANDING, RECOVERY, OR RE-ENTRY	NOTES
Kosmos 403	2 Apr 1971	14 Apr 1971	
Kosmos 406	14 Apr 1971	24 Apr 1971	
Kosmos 410	6 May 1971	18 May 1971	
Kosmos 420	18 May 1971	29 May 1971	
Kosmos 424	28 May 1971	10 Jun 1971	
Kosmos 427	11 Jun 1971	23 Jun 1971	
Kosmos 428	24 Jun 1971	6 Jul 1971	
Kosmos 429	20 Jul 1971	2 Aug 1971	
Kosmos 430	23 Jul 1971	5 Aug 1971	
Kosmos 431	30 Jul 1971	11 Aug 1971	
Kosmos 432	5 Aug 1971	18 Aug 1971	
Kosmos 438	14 Sep 1971	27 Sep 1971	
Kosmos 439	21 Sep 1971	2 Oct 1971	
Kosmos 441	28 Sep 1971	10 Oct 1971	
Kosmos 442	29 Sep 1971	12 Oct 1971	
Kosmos 443	7 Oct 1971	19 Oct 1971	
Kosmos 452	14 Oct 1971	27 Oct 1971	
Kosmos 454	2 Nov 1971	16 Nov 1971	
Kosmos 456	19 Nov 1971	2 Dec 1971	
Kosmos 463	6 Dec 1971	11 Dec 1971	
Kosmos 464	10 Dec 1971	16 Dec 1971	
Kosmos 466	16 Dec 1971	27 Dec 1971	
Kosmos 470	27 Dec 1971	6 Jan 1972	
Kosmos 471	12 Jan 1972	25 Jan 1972	
Kosmos 473	3 Feb 1972	15 Feb 1972	
Kosmos 474	16 Feb 1972	29 Feb 1972	
Kosmos 477	4 Mar 1972	16 Mar 1972	
Kosmos 478	15 Mar 1972	28 Mar 1972	
Kosmos 483	3 Apr 1972	15 Apr 1972	
Kosmos 484	6 Apr 1972	18 Apr 1972	
Kosmos 486	14 Apr 1972	27 Apr 1972	
Kosmos 488	5 May 1972	18 May 1972	
Kosmos 490	17 May 1972	29 May 1972	
Kosmos 491	25 May 1972	8 Jun 1972	
Kosmos 492	9 Jun 1972	22 Jun 1972	
Kosmos 493	21 Jun 1972	3 Jul 1972	
Kosmos 495	23 Jun 1972	6 Jul 1972	
Kosmos 499	6 Jul 1972	17 Jul 1972	
Kosmos 502	13 Jul 1972	25 Jul 1972	
Kosmos 503	19 Jul 1972	1 Aug 1972	
Kosmos 512	28 Jul 1972	9 Aug 1972	
Kosmos 513	2 Aug 1972	15 Aug 1972	
Kosmos 515	18 Aug 1972	31 Aug 1972	
Kosmos 517	30 Aug 1972	11 Sep 1972	
Kosmos 518	15 Sep 1972	24 Sep 1972	
Kosmos 519	16 Sep 1972	26 Sep 1972	
Kosmos 522	4 Oct 1972	17 Oct 1972	
Kosmos 525	18 Oct 1972	29 Oct 1972	

TABLE 1　59

MISSION	LAUNCH	LANDING, RECOVERY, OR RE-ENTRY	NOTES
Kosmos 527	31 Oct 1972	13 Nov 1972	
Kosmos 537	25 Nov 1972	7 Dec 1972	
Kosmos 538	14 Dec 1972	27 Dec 1972	
Kosmos 541	27 Dec 1972	8 Jan 1973	
Kosmos 543	11 Jan 1973	24 Jan 1973	
Kosmos 547	1 Feb 1973	13 Feb 1973	
Kosmos 548	8 Feb 1973	21 Feb 1973	
Kosmos 550	1 Mar 1973	11 Mar 1973	
Kosmos 551	6 Mar 1973	20 Mar 1973	
Kosmos 552	22 Mar 1973	3 Apr 1973	
Kosmos 554	19 Apr 1973	27 May 1973	
Kosmos 555	25 Apr 1973	7 May 1973	
Kosmos 556	5 May 1973	14 May 1973	
Kosmos 559	18 May 1973	23 May 1973	
Kosmos 560	23 May 1973	5 Jun 1973	
Kosmos 561	25 May 1973	6 Jun 1973	
Kosmos 563	6 Jun 1973	18 Jun 1973	
Kosmos 572	10 Jun 1973	23 Jun 1973	
Kosmos 575	21 Jun 1973	3 Jul 1973	
Kosmos 576	27 Jun 1973	9 Jul 1973	
Kosmos 577	25 Jul 1973	7 Aug 1973	
Kosmos 578	1 Aug 1973	13 Aug 1973	
Kosmos 579	21 Aug 1973	3 Sep 1973	
Kosmos 581	24 Aug 1973	6 Sep 1973	
Kosmos 583	30 Aug 1973	12 Sep 1973	
Kosmos 584	6 Sep 1973	20 Sep 1973	
Kosmos 587	21 Sep 1973	4 Oct 1973	
Kosmos 596	3 Oct 1973	9 Oct 1973	
Kosmos 597	6 Oct 1973	12 Oct 1973	
Kosmos 598	10 Oct 1973	16 Oct 1973	
Kosmos 599	15 Oct 1973	28 Oct 1973	
Kosmos 600	16 Oct 1973	23 Oct 1973	
Kosmos 602	20 Oct 1973	29 Oct 1973	
Kosmos 603	27 Oct 1973	9 Nov 1973	
Kosmos 607	10 Nov 1973	22 Nov 1973	
Kosmos 609	21 Nov 1973	4 Dec 1973	
Kosmos 612	28 Nov 1973	11 Dec 1973	
Kosmos 616	17 Dec 1973	28 Dec 1973	
Kosmos 625	21 Dec 1973	3 Jan 1974	
Kosmos 629	24 Jan 1974	5 Feb 1974	
Kosmos 630	30 Jan 1974	13 Feb 1974	
Kosmos 632	12 Feb 1974	26 Feb 1974	
Kosmos 635	14 Mar 1974	26 Mar 1974	
Kosmos 636	20 Mar 1974	3 Apr 1974	
Kosmos 639	4 Apr 1974	15 Apr 1974	
Kosmos 640	11 Apr 1974	23 Apr 1974	
Kosmos 649	29 Apr 1974	11 May 1974	
Kosmos 652	15 May 1974	23 May 1974	

MISSION	LAUNCH	LANDING, RECOVERY, OR RE-ENTRY	NOTES
Kosmos 653	15 May 1974	27 May 1974	
Kosmos 657	30 May 1974	13 Jun 1974	
Kosmos 658	6 Jun 1974	18 Jun 1974	
Kosmos 659	13 Jun 1974	26 Jun 1974	
Kosmos 664	29 Jun 1974	11 Jul 1974	
Kosmos 666	12 Jul 1974	25 Jul 1974	
Kosmos 667	25 Jul 1974	7 Aug 1974	
Kosmos 669	26 Jul 1974	8 Aug 1974	
Kosmos 671	7 Aug 1974	20 Aug 1974	
Kosmos 674	29 Aug 1974	7 Sep 1974	
Kosmos 685	20 Sep 1974	2 Oct 1974	
Kosmos 688	18 Oct 1974	30 Oct 1974	
Kosmos 691	25 Oct 1974	6 Nov 1974	
Kosmos 692	1 Nov 1974	13 Nov 1974	
Kosmos 693	4 Nov 1974	16 Nov 1974	
Kosmos 694	16 Nov 1974	29 Nov 1974	
Kosmos 696	27 Nov 1974	9 Dec 1974	
Kosmos 697	13 Dec 1974	25 Dec 1974	
Kosmos 701	27 Dec 1974	9 Jan 1975	
Kosmos 702	17 Jan 1975	29 Jan 1975	
Kosmos 704	23 Jan 1975	6 Feb 1975	
Kosmos 709	12 Feb 1975	25 Feb 1975	
Kosmos 710	26 Feb 1975	12 Mar 1975	
Kosmos 719	12 Mar 1975	25 Mar 1975	
Kosmos 720	21 Mar 1975	1 Apr 1975	
Kosmos 721	26 Mar 1975	7 Apr 1975	
Kosmos 722	27 Mar 1975	9 Apr 1975	
Kosmos 727	16 Apr 1975	28 Apr 1975	
Kosmos 728	18 Apr 1975	29 Apr 1975	
Kosmos 730	24 Apr 1975	6 May 1975	
Kosmos 731	21 May 1975	2 Jun 1975	
Kosmos 740	28 May 1975	10 Jun 1975	
Kosmos 741	30 May 1975	11 Jun 1975	
Kosmos 742	3 Jun 1975	15 Jun 1975	
Kosmos 743	12 Jun 1975	25 Jun 1975	
Kosmos 746	25 Jun 1975	8 Jul 1975	
Kosmos 747	27 Jun 1975	9 Jul 1975	
Kosmos 748	3 Jul 1975	16 Jul 1975	
Kosmos 751	23 Jul 1975	4 Aug 1975	
Kosmos 753	31 Jul 1975	13 Aug 1975	
Kosmos 754	13 Aug 1975	26 Aug 1975	
Kosmos 757	27 Aug 1975	9 Sep 1975	
Kosmos 758	5 Sep 1975	25 Sep 1975	
Kosmos 759	12 Sep 1975	23 Sep 1975	
Kosmos 760	16 Sep 1975	30 Sep 1975	
Kosmos 769	23 Sep 1975	5 Oct 1975	
Kosmos 771	25 Sep 1975	8 Oct 1975	
Kosmos 774	1 Oct 1975	15 Oct 1975	

TABLE 1   61

MISSION	LAUNCH	LANDING, RECOVERY, OR RE-ENTRY	NOTES
Kosmos 776	17 Oct 1975	29 Oct 1975	
Kosmos 779	4 Nov 1975	18 Nov 1975	
Kosmos 780	21 Nov 1975	3 Dec 1975	
Kosmos 784	3 Dec 1975	15 Dec 1975	
Kosmos 786	16 Dec 1975	29 Dec 1975	
Kosmos 788	7 Jan 1976	20 Jan 1976	
Kosmos 799	29 Jan 1976	10 Feb 1976	
Kosmos 802	11 Feb 1976	25 Feb 1976	
Kosmos 805	20 Feb 1976	11 Mar 1976	
Kosmos 806	10 Mar 1976	23 Mar 1976	
Kosmos 809	18 Mar 1976	30 Mar 1976	
Kosmos 810	26 Mar 1976	8 Apr 1976	
Kosmos 811	31 Mar 1976	12 Apr 1976	
Kosmos 813	9 Apr 1976	21 Apr 1976	
Kosmos 815	28 Apr 1976	11 May 1976	
Kosmos 817	5 May 1976	18 May 1976	
Kosmos 819	20 May 1976	1 Jun 1976	
Kosmos 820	21 May 1976	2 Jun 1976	
Kosmos 821	26 May 1976	8 Jun 1976	
Kosmos 824	8 Jun 1976	21 Jun 1976	
Kosmos 833	16 Jun 1976	29 Jun 1976	
Kosmos 834	24 Jun 1976	6 Jul 1976	
Kosmos 835	29 Jun 1976	12 Jul 1976	
Kosmos 840	14 Jul 1976	26 Jul 1976	
Kosmos 844	22 Jul 1976	30 Aug 1976	
Kosmos 847	4 Aug 1976	17 Aug 1976	
Kosmos 848	12 Aug 1976	25 Aug 1976	
Kosmos 852	28 Aug 1976	10 Sep 1976	
Kosmos 854	3 Sep 1976	16 Sep 1976	
Kosmos 855	21 Sep 1976	3 Oct 1976	
Kosmos 856	22 Sep 1976	5 Oct 1976	
Kosmos 857	24 Sep 1976	7 Oct 1976	
Kosmos 859	10 Oct 1976	21 Oct 1976	
Kosmos 863	22 Oct 1976	1 Nov 1976	
Kosmos 865	1 Nov 1976	13 Nov 1976	
Kosmos 866	11 Nov 1976	23 Nov 1976	
Kosmos 867	23 Nov 1976	6 Dec 1976	
Kosmos 879	9 Dec 1976	22 Dec 1976	
Kosmos 884	17 Dec 1976	29 Dec 1976	
Kosmos 888	6 Jan 1977	19 Jan 1977	
Kosmos 889	20 Jan 1977	1 Feb 1977	
Kosmos 892	9 Feb 1977	22 Feb 1977	
Kosmos 896	3 Mar 1977	16 Mar 1977	
Kosmos 897	10 Mar 1977	23 Mar 1977	
Kosmos 898	17 Mar 1977	30 Mar 1977	
Kosmos 902	7 Apr 1977	20 Apr 1977	
Kosmos 904	20 Apr 1977	4 May 1977	
Kosmos 905	26 Apr 1977	26 May 1977	

MISSION	LAUNCH	LANDING, RECOVERY, OR RE-ENTRY	NOTES
Kosmos 907	5 May 1977	16 May 1977	
Kosmos 908	17 May 1977	31 May 1977	
Kosmos 912	26 May 1977	8 Jun 1977	
Kosmos 914	31 May 1977	13 Jun 1977	
Kosmos 915	8 Jun 1977	21 Jun 1977	
Kosmos 916	10 Jun 1977	21 Jun 1977	
Kosmos 920	22 Jun 1977	5 Jul 1977	
Kosmos 922	30 Jun 1977	13 Jul 1977	
Kosmos 927	12 Jul 1977	25 Jul 1977	
Kosmos 932	20 Jul 1977	2 Aug 1977	
Kosmos 934	27 Jul 1977	9 Aug 1977	
Kosmos 935	29 Jul 1977	11 Aug 1977	
Kosmos 938	24 Aug 1977	6 Sep 1977	
Kosmos 947	27 Aug 1977	9 Sep 1977	
Kosmos 948	2 Sep 1977	15 Sep 1977	
Kosmos 949	6 Sep 1977	6 Oct 1977	
Kosmos 950	13 Sep 1977	27 Sep 1977	
Kosmos 953	16 Sep 1977	29 Sep 1977	
Kosmos 957	30 Sep 1977	13 Oct 1977	
Kosmos 958	11 Oct 1977	24 Oct 1977	
Kosmos 964	4 Dec 1977	17 Dec 1977	
Kosmos 966	12 Dec 1977	24 Dec 1977	
Kosmos 969	20 Dec 1977	3 Jan 1978	
Kosmos 973	27 Dec 1977	9 Jan 1978	
Kosmos 974	6 Jan 1978	19 Jan 1978	
Kosmos 984	13 Jan 1978	26 Jan 1978	
Kosmos 986	24 Jan 1978	7 Feb 1978	
Kosmos 987	31 Jan 1978	14 Feb 1978	
Kosmos 988	8 Feb 1978	20 Feb 1978	
Kosmos 989	14 Feb 1978	28 Feb 1978	
Kosmos 992	4 Mar 1978	17 Mar 1978	
Kosmos 993	10 Mar 1978	23 Mar 1978	
Kosmos 995	17 Mar 1978	30 Mar 1978	
Kosmos 999	30 Mar 1978	12 Apr 1978	
Kosmos 1002	6 Apr 1978	19 Apr 1978	
Kosmos 1003	20 Apr 1978	4 May 1978	
Kosmos 1004	5 May 1978	18 May 1978	
Kosmos 1007	16 May 1978	29 May 1978	
Kosmos 1010	23 May 1978	5 Jun 1978	
Kosmos 1012	25 May 1978	7 Jun 1978	
Kosmos 1021	10 Jun 1978	23 Jun 1978	
Kosmos 1022	12 Jun 1978	25 Jun 1978	
Kosmos 1026	2 Jul 1978	6 Jul 1978	
Kosmos 1028	5 Aug 1978	4 Sep 1978	
Kosmos 1029	29 Aug 1978	8 Sep 1978	
Kosmos 1031	9 Sep 1978	22 Sep 1978	
Kosmos 1032	19 Sep 1978	2 Oct 1978	
Kosmos 1033	3 Oct 1978	16 Oct 1978	

TABLE 1   63

MISSION	LAUNCH	LANDING, RECOVERY, OR RE-ENTRY	NOTES
Kosmos 1042	6 Oct 1978	19 Oct 1978	
Kosmos 1044	17 Oct 1978	30 Oct 1978	
Kosmos 1046	1 Nov 1978	13 Nov 1978	
Kosmos 1047	15 Nov 1978	28 Nov 1978	
Kosmos 1049	21 Nov 1978	4 Dec 1978	
Kosmos 1050	28 Nov 1978	12 Dec 1978	
Kosmos 1059	7 Dec 1978	20 Dec 1978	
Kosmos 1060	8 Dec 1978	21 Dec 1978	
Kosmos 1061	14 Dec 1978	27 Dec 1978	
Kosmos 1068	26 Dec 1978	8 Jan 1979	
Kosmos 1069	28 Dec 1978	10 Jan 1979	
Kosmos 1070	11 Jan 1979	20 Jan 1979	
Kosmos 1071	13 Jan 1979	26 Jan 1979	
Kosmos 1073	30 Jan 1979	12 Feb 1979	
Kosmos 1078	22 Feb 1979	2 Mar 1979	
Kosmos 1079	27 Feb 1979	11 Mar 1979	
Kosmos 1080	14 Mar 1979	28 Mar 1979	
Kosmos 1090	31 Mar 1979	13 Apr 1979	
Kosmos 1095	20 Apr 1979	3 Apr 1979	
Kosmos 1097	27 Apr 1979	27 May 1979	
Kosmos 1098	15 May 1979	28 May 1979	
Kosmos 1099	17 May 1979	30 May 1979	
Kosmos 1102	25 May 1979	7 Jun 1979	
Kosmos 1103	31 May 1979	14 Jun 1979	
Kosmos 1105	8 Jun 1979	21 Jun 1979	
Kosmos 1106	12 Jun 1979	25 Jun 1979	
Kosmos 1107	15 Jun 1979	29 Jun 1979	
Kosmos 1108	22 Jun 1979	5 Jul 1979	
Kosmos 1111	29 Jun 1979	14 Jul 1979	
Kosmos 1113	10 Jul 1979	23 Jul 1979	
Kosmos 1115	13 Jul 1979	26 Jul 1979	
Kosmos 1117	25 Jul 1979	7 Aug 1979	
Kosmos 1118	27 Jul 1979	9 Aug 1979	
Kosmos 1119	3 Aug 1979	18 Aug 1979	
Kosmos 1120	11 Aug 1979	24 Aug 1979	
Kosmos 1121	14 Aug 1979	13 Sep 1979	
Kosmos 1122	17 Aug 1979	30 Aug 1979	
Kosmos 1123	21 Aug 1979	3 Sep 1979	
Kosmos 1126	31 Aug 1979	14 Sep 1979	
Kosmos 1127	5 Sep 1979	18 Sep 1979	
Kosmos 1128	14 Sep 1979	27 Sep 1979	
Kosmos 1138	28 Sep 1979	11 Oct 1979	
Kosmos 1139	5 Oct 1979	18 Oct 1979	
Kosmos 1142	20 Oct 1979	4 Nov 1979	
Kosmos 1144	2 Nov 1979	4 Dec 1979	
Kosmos 1147	12 Dec 1979	26 Dec 1979	
Kosmos 1148	28 Dec 1979	10 Jan 1980	
Kosmos 1149	9 Jan 1980	23 Jan 1980	

MISSION	LAUNCH	LANDING, RECOVERY, OR RE-ENTRY	NOTES
Kosmos 1152	24 Jan 1980	6 Feb 1980	
Kosmos 1155	7 Feb 1980	21 Feb 1980	
Kosmos 1165	21 Feb 1980	5 Mar 1980	
Kosmos 1166	4 Mar 1980	18 Mar 1980	
Kosmos 1170	1 Apr 1980	12 Apr 1980	
Kosmos 1173	17 Apr 1980	28 Apr 1980	
Kosmos 1177	29 Apr 1980	12 Jun 1980	
Kosmos 1178	7 May 1980	22 May 1980	
Kosmos 1180	15 May 1980	26 May 1980	
Kosmos 1182	23 May 1980	5 Jun 1980	
Kosmos 1183	28 May 1980	11 Jun 1980	
Kosmos 1185	6 Jun 1980	20 Jun 1980	
Kosmos 1187	12 Jun 1980	26 Jun 1980	
Kosmos 1189	26 Jun 1980	10 Jul 1980	
Kosmos 1200	9 Jul 1980	23 Jul 1980	
Kosmos 1201	15 Jul 1980	28 Jul 1980	
Kosmos 1202	24 Jul 1980	7 Aug 1980	
Kosmos 1203	31 Jul 1980	14 Aug 1980	
Kosmos 1205	12 Aug 1980	26 Aug 1980	
Kosmos 1207	22 Aug 1980	8 Sep 1980	
Kosmos 1208	26 Aug 1980	24 Sep 1980	
Kosmos 1209	3 Sep 1980	17 Sep 1980	
Kosmos 1210	9 Sep 1980	3 Oct 1980	
Kosmos 1211	23 Sep 1980	4 Oct 1980	
Kosmos 1212	26 Sep 1980	9 Oct 1980	
Kosmos 1213	3 Oct 1980	17 Oct 1980	
Kosmos 1214	10 Oct 1980	23 Oct 1980	
Kosmos 1216	16 Oct 1980	30 Oct 1980	
Kosmos 1218	30 Oct 1980	12 Dec 1980	
Kosmos 1219	31 Oct 1980	13 Nov 1980	
Kosmos 1221	12 Nov 1980	26 Nov 1980	
Kosmos 1224	1 Dec 1980	15 Dec 1980	
Kosmos 1227	16 Dec 1980	28 Dec 1980	
Kosmos 1236	26 Dec 1980	21 Jan 1981	
Kosmos 1237	6 Jan 1981	21 Jan 1981	
Kosmos 1239	16 Jan 1981	28 Jan 1981	
Kosmos 1240	20 Jan 1981	17 Feb 1981	
Kosmos 1245	13 Feb 1981	27 Feb 1981	
Kosmos 1246	18 Feb 1981	13 Mar 1981	
Kosmos 1248	5 Mar 1981	4 Apr 1981	
Kosmos 1259	17 Mar 1981	31 Mar 1981	
Kosmos 1262	7 Apr 1981	21 Apr 1981	
Kosmos 1264	15 Apr 1981	19 Apr 1981	
Kosmos 1265	16 Apr 1981	28 Apr 1981	
Kosmos 1268	28 Apr 1981	12 May 1981	
Kosmos 1270	18 May 1981	17 Jun 1981	
Kosmos 1272	21 May 1981	4 Jun 1981	
Kosmos 1274	3 Jun 1981	3 Jul 1981	

TABLE 1   65

MISSION	LAUNCH	LANDING, RECOVERY, OR RE-ENTRY	NOTES
Kosmos 1277	17 Jun 1981	1 Jul 1981	
Kosmos 1279	1 Jul 1981	15 Jul 1981	
Kosmos 1281	7 Jul 1981	21 Jul 1981	
Kosmos 1282	15 Jul 1981	14 Aug 1981	
Kosmos 1296	13 Aug 1981	13 Sep 1981	
Kosmos 1297	18 Aug 1981	25 Aug 1981	
Kosmos 1298	21 Aug 1981	2 Oct 1981	
Kosmos 1303	4 Sep 1981	18 Sep 1981	
Kosmos 1307	15 Sep 1981	28 Sep 1981	
Kosmos 1309	18 Sep 1981	1 Oct 1981	
Kosmos 1313	1 Oct 1981	15 Oct 1981	
Kosmos 1316	15 Oct 1981	29 Oct 1981	
Kosmos 1318	3 Nov 1981	4 Dec 1981	
Kosmos 1319	13 Nov 1981	27 Nov 1981	
Kosmos 1329	4 Dec 1981	18 Dec 1981	
Kosmos 1330	19 Dec 1981	19 Jan 1982	
Kosmos 1332	12 Jan 1982	25 Jan 1982	
Kosmos 1334	20 Jan 1982	3 Feb 1982	
Kosmos 1336	30 Jan 1982	26 Feb 1982	
Kosmos 1338	16 Feb 1982	2 Mar 1982	
Kosmos 1342	5 Mar 1982	19 Mar 1982	
Kosmos 1343	17 Mar 1982	31 Mar 1982	
Kosmos 1347	2 Apr 1982	21 May 1982	
Kosmos 1350	15 Apr 1982	16 May 1982	
Kosmos 1352	21 Apr 1982	5 May 1982	
Kosmos 1353	23 Apr 1982	6 May 1982	
Kosmos 1368	21 May 1982	3 Jun 1982	
Kosmos 1369	25 May 1982	8 Jun 1982	
Kosmos 1370	28 May 1982	11 Jul 1982	
Kosmos 1373	2 Jun 1982	16 Jun 1982	
Kosmos 1376	8 Jun 1982	22 Jun 1982	
Kosmos 1377	8 Jun 1982	22 Jul 1982	
Kosmos 1381	18 Jun 1982	1 Jul 1982	
Kosmos 1384	30 Jun 1982	30 Jul 1982	
Kosmos 1385	6 Jul 1982	20 Jul 1982	
Kosmos 1387	13 Jul 1982	26 Jul 1982	
Kosmos 1396	27 Jul 1982	10 Aug 1982	
Kosmos 1398	3 Aug 1982	13 Aug 1982	
Kosmos 1399	4 Aug 1982	16 Sep 1982	
Kosmos 1401	20 Aug 1982	3 Sep 1982	
Kosmos 1403	1 Sep 1982	15 Sep 1982	
Kosmos 1404	1 Sep 1982	15 Sep 1982	
Kosmos 1406	8 Sep 1982	21 Sep 1982	
Kosmos 1407	15 Sep 1982	16 Oct 1982	
Kosmos 1411	30 Sep 1982	14 Oct 1982	
Kosmos 1416	14 Oct 1982	28 Oct 1982	
Kosmos 1419	2 Nov 1982	16 Nov 1982	
Kosmos 1421	18 Nov 1982	2 Dec 1982	

MISSION	LAUNCH	LANDING, RECOVERY, OR RE-ENTRY	NOTES
Kosmos 1422	3 Dec 1982	17 Dec 1982	
Kosmos 1424	16 Dec 1982	28 Jan 1983	
Kosmos 1425	23 Dec 1982	6 Jan 1983	
Kosmos 1426	28 Dec 1982	5 Mar 1983	
Kosmos 1438	27 Jan 1983	7 Feb 1983	
Kosmos 1439	6 Feb 1983	22 Feb 1983	
Kosmos 1440	10 Feb 1983	24 Feb 1983	
Kosmos 1442	25 Feb 1983	11 Apr 1983	
Kosmos 1444	2 Mar 1983	16 Mar 1983	
Kosmos 1446	16 Mar 1983	30 Mar 1983	
Kosmos 1449	31 Mar 1983	15 Apr 1983	
Kosmos 1451	8 Apr 1983	22 Apr 1983	
Kosmos 1454	22 Apr 1983	22 May 1983	
Kosmos 1457	26 Apr 1983	8 Jun 1983	
Kosmos 1458	28 Apr 1983	11 May 1983	
Kosmos 1460	6 May 1983	20 May 1983	
Kosmos 1462	17 May 1983	31 May 1983	
Kosmos 1466	26 May 1983	6 Jul 1983	
Kosmos 1467	31 May 1983	12 Jun 1983	
Kosmos 1468	7 Jun 1983	21 Jun 1983	
Kosmos 1469	14 Jun 1983	24 Jun 1983	
Kosmos 1471	28 Jun 1983	28 Jul 1983	
Kosmos 1472	5 Jul 1983	19 Jul 1983	
Kosmos 1482	13 Jul 1983	27 Jul 1983	
Kosmos 1483	20 Jul 1983	3 Aug 1983	
Kosmos 1485	26 Jul 1983	9 Aug 1983	
Kosmos 1487	5 Aug 1983	19 Aug 1983	
Kosmos 1488	9 Aug 1983	23 Aug 1983	
Kosmos 1489	10 Aug 1983	23 Sep 1983	
Kosmos 1493	23 Aug 1983	6 Sep 1983	
Kosmos 1495	3 Sep 1983	16 Sep 1983	
Kosmos 1496	7 Sep 1983	19 Oct 1983	
Kosmos 1497	9 Sep 1983	23 Sep 1983	
Kosmos 1498	14 Sep 1983	28 Sep 1983	
Kosmos 1499	17 Sep 1983	1 Oct 1983	
Kosmos 1504	14 Oct 1983	6 Dec 1983	
Kosmos 1505	21 Oct 1983	4 Nov 1983	
Kosmos 1509	17 Nov 1983	3 Dec 1983	
Kosmos 1511	30 Nov 1983	13 Jan 1984	
Kosmos 1512	7 Dec 1983	21 Dec 1983	
Kosmos 1516	27 Dec 1983	9 Feb 1984	
Kosmos 1530	11 Jan 1984	25 Jan 1984	
Kosmos 1532	14 Jan 1984	26 Feb 1984	
Kosmos 1533	26 Jan 1984	9 Feb 1984	
Kosmos 1537	16 Feb 1984	1 Mar 1984	
Kosmos 1539	28 Feb 1984	9 Apr 1984	
Kosmos 1542	7 Mar 1984	21 Mar 1984	
Kosmos 1543	10 Mar 1984	5 Apr 1984	

TABLE 1 67

MISSION	LAUNCH	LANDING, RECOVERY, OR RE-ENTRY	NOTES
Kosmos 1545	21 Mar 1984	5 Apr 1984	
Kosmos 1548	10 Apr 1984	25 May 1984	
Kosmos 1549	19 Apr 1984	3 May 1984	
Kosmos 1551	11 May 1984	23 May 1984	
Kosmos 1552	14 May 1984	3 Nov 1984	
Kosmos 1558	25 May 1984	8 Jul 1984	
Kosmos 1568	1 Jun 1984	14 Jun 1984	
Kosmos 1571	11 Jun 1984	26 Jun 1984	
Kosmos 1573	19 Jun 1984	28 Jun 1984	
Kosmos 1576	26 Jun 1984	24 Aug 1984	
Kosmos 1580	29 Jun 1984	13 Jul 1984	
Kosmos 1582	19 Jul 1984	2 Aug 1984	
Kosmos 1583	24 Jul 1984	8 Aug 1984	
Kosmos 1585	31 Jul 1984	28 Sep 1984	
Kosmos 1587	6 Aug 1984	31 Aug 1984	
Kosmos 1592	4 Sep 1984	18 Sep 1984	
Kosmos 1599	25 Sep 1984	20 Nov 1984	
Kosmos 1600	27 Sep 1984	11 Oct 1984	
Kosmos 1608	14 Nov 1984	17 Dec 1984	
Kosmos 1609	14 Nov 1984	28 Nov 1984	
Kosmos 1611	21 Nov 1984	11 Jan 1985	
Kosmos 1613	29 Nov 1984	24 Dec 1984	
Kosmos 1616	9 Jan 1985	4 Mar 1985	
Kosmos 1623	16 Jan 1985	30 Jan 1985	
Kosmos 1628	6 Feb 1985	20 Feb 1985	
Kosmos 1630	27 Feb 1985	23 Apr 1985	
Kosmos 1632	1 Mar 1985	15 Mar 1985	
Kosmos 1643	25 Mar 1985	18 Oct 1985	
Kosmos 1644	3 Apr 1985	17 Apr 1985	
Kosmos 1645	16 Apr 1985	29 Apr 1985	
Kosmos 1647	19 Apr 1985	11 Jun 1985	
Kosmos 1648	25 Apr 1985	6 May 1985	
Kosmos 1649	15 May 1985	29 May 1985	
Kosmos 1654	23 May 1985	7 Aug 1985	
Kosmos 1659	13 Jun 1985	27 Jun 1985	
Kosmos 1664	26 Jun 1985	5 Jul 1985	
Kosmos 1665	3 Jul 1985	17 Jul 1985	
Kosmos 1668	15 Jul 1985	29 Jul 1985	
Kosmos 1671	2 Aug 1985	16 Aug 1985	
Kosmos 1673	8 Aug 1985	19 Sep 1985	
Kosmos 1676	16 Aug 1985	14 Oct 1985	
Kosmos 1679	29 Aug 1985	18 Oct 1985	
Kosmos 1683	19 Sep 1985	4 Oct 1985	
Kosmos 1685	27 Sep 1985	10 Oct 1985	
Kosmos 1696	16 Oct 1985	30 Oct 1985	
Kosmos 1699	25 Oct 1985	23 Dec 1985	
Kosmos 1702	13 Nov 1985	27 Nov 1985	
Kosmos 1705	3 Dec 1985	17 Dec 1985	

MISSION	LAUNCH	LANDING, RECOVERY, OR RE-ENTRY	NOTES
Kosmos 1706	11 Dec 1985	9 Feb 1986	
Kosmos 1713	27 Dec 1985	22 Jan 1986	
Kosmos 1715	8 Jan 1986	22 Jan 1986	
Kosmos 1724	15 Jan 1986	15 Mar 1986	
Kosmos 1728	28 Jan 1986	11 Feb 1986	
Kosmos 1730	4 Feb 1986	13 Feb 1986	
Kosmos 1731	7 Feb 1986	3 Oct 1986	
Kosmos 1734	26 Feb 1986	26 Apr 1986	
Kosmos 1739	9 Apr 1986	7 Jun 1986	
Kosmos 1740	15 Apr 1986	28 Apr 1986	
Kosmos 1742	14 May 1986	28 May 1986	
Kosmos 1744	21 May 1986	4 Jun 1986	
Kosmos 1746	28 May 1986	12 Jun 1986	
Kosmos 1747	29 May 1986	12 Jun 1986	
Kosmos 1756	6 Jun 1986	4 Aug 1986	
Kosmos 1757	11 Jun 1986	25 Jun 1986	
Kosmos 1760	19 Jun 1986	3 Jul 1986	
Kosmos 1764	17 Jul 1986	11 Sep 1986	
Kosmos 1765	24 Jul 1986	7 Aug 1986	
Kosmos 1767	30 Jul 1986	16 Aug 1986	
Kosmos 1768	2 Aug 1986	16 Aug 1986	
Kosmos 1770	6 Aug 1986	2 Feb 1987	
Kosmos 1772	21 Aug 1986	3 Sep 1986	
Kosmos 1773	27 Aug 1986	21 Oct 1986	
Kosmos 1775	3 Sep 1986	17 Sep 1986	
Kosmos 1781	17 Sep 1986	1 Oct 1986	
Kosmos 1784	6 Oct 1986	11 Nov 1986	
Kosmos 1787	22 Oct 1986	4 Nov 1986	
Kosmos 1789	31 Oct 1986	14 Nov 1986	
Kosmos 1790	4 Nov 1986	18 Nov 1986	
Kosmos 1792	13 Nov 1986	5 Jan 1987	
Kosmos 1804	4 Dec 1986	18 Dec 1986	
Kosmos 1807	16 Dec 1986	23 Jan 1987	
Kosmos 1810	26 Dec 1986	11 Sep 1987	
Kosmos 1811	9 Jan 1987	13 Feb 1987	
Kosmos 1813	15 Jan 1987	29 Jan 1987	
Kosmos 1819	7 Feb 1987	18 Feb 1987	
Kosmos 1822	19 Feb 1987	5 Mar 1987	
Kosmos 1824	26 Feb 1987	22 Apr 1987	
Kosmos 1826	11 Mar 1987	25 Mar 1987	
Kosmos 1835	9 Apr 1987	4 Jun 1987	
Kosmos 1836	16 Apr 1987	2 Dec 1987	
Kosmos 1837	22 Apr 1987	28 Apr 1987	
Kosmos 1843	5 May 1987	19 May 1987	
Kosmos 1845	13 May 1987	27 May 1987	
Kosmos 1846	21 May 1987	4 Jun 1987	
Kosmos 1847	26 May 1987	22 Jul 1987	
Kosmos 1848	28 May 1987	11 Jun 1987	

TABLE 1   69

MISSION	LAUNCH	LANDING, RECOVERY, OR RE-ENTRY	NOTES
Kosmos 1863	4 Jul 1987	18 Jul 1987	
Kosmos 1865	8 Jul 1987	4 Aug 1987	
Kosmos 1866	9 Jul 1987	6 Nov 1987	
Kosmos 1872	19 Aug 1987	30 Aug 1987	
Kosmos 1874	3 Sep 1987	17 Sep 1987	
Kosmos 1881	11 Sep 1987		
Kosmos 1886	17 Sep 1987	2 Nov 1987	
Kosmos 1889	9 Oct 1987	23 Oct 1987	
Kosmos 1893	22 Oct 1987	16 Dec 1987	
Kosmos 1895	11 Nov 1987	26 Nov 1987	
Kosmos 1896	14 Nov 1987	25 Dec 1987	
Kosmos 1899	7 Dec 1987	21 Dec 1987	
Kosmos 1901	14 Dec 1987		
Kosmos 1902	15 Dec 1987		
Kosmos 1905	25 Dec 1987		
Kosmos 1907	29 Dec 1987		

KOSMOS (Group Launches)

MISSION	LAUNCH		NOTES
Kosmos 336	25 Apr 1970		Grouped up to Kosmos 343 *
Kosmos 411	7 May 1971		Grouped up to Kosmos 418
Kosmos 444	13 Oct 1971		Grouped up to Kosmos 451
Kosmos 504	20 Jul 1972		Grouped up to Kosmos 511
Kosmos 528	1 Nov 1972		Grouped up to Kosmos 535
Kosmos 564	8 Jun 1973		Grouped up to Kosmos 571
Kosmos 588	3 Oct 1973		Grouped up to Kosmos 595
Kosmos 617	19 Dec 1973		Grouped up to Kosmos 624
Kosmos 641	23 Apr 1974		Grouped up to Kosmos 648
Kosmos 677	19 Sep 1974		Grouped up to Kosmos 684
Kosmos 711	28 Feb 1975		Grouped up to Kosmos 718
Kosmos 732	28 May 1975		Grouped up to Kosmos 739
Kosmos 761	17 Sep 1975		Grouped up to Kosmos 768
Kosmos 791	28 Jan 1976		Grouped up to Kosmos 798
Kosmos 825	15 Jun 1976		Grouped up to Kosmos 832
Kosmos 871	7 Dec 1976		Grouped up to Kosmos 878
Kosmos 939	24 Aug 1977		Grouped up to Kosmos 946
Kosmos 976	10 Jan 1978		Grouped up to Kosmos 983
Kosmos 1013	7 Jun 1978		Grouped up to Kosmos 1020
Kosmos 1034	4 Oct 1978		Grouped up to Kosmos 1041
Kosmos 1051	5 Dec 1978		Grouped up to Kosmos 1058
Kosmos 1081	15 Mar 1979		Grouped up to Kosmos 1088
Kosmos 1130	25 Sep 1979		Grouped up to Kosmos 1137
Kosmos 1156	11 Feb 1980		Grouped up to Kosmos 1163
Kosmos 1192	9 Jul 1980		Grouped up to Kosmos 1199
Kosmos 1228	23 Dec 1980		Grouped up to Kosmos 1235
Kosmos 1250	6 Mar 1981		Grouped up to Kosmos 1257
Kosmos 1287	6 Aug 1981		Grouped up to Kosmos 1294

* **"Grouped up" means that Kosmos 336–343 were all launched aboard a single rocket. Similarly for all subsequent groups.**

MISSION	LAUNCH	LANDING, RECOVERY, OR RE-ENTRY	NOTES
Kosmos 1320	28 Nov 1981		Grouped up to Kosmos 1327
Kosmos 1357	6 May 1982		Grouped up to Kosmos 1364
Kosmos 1388	21 Jul 1982		Grouped up to Kosmos 1395
Kosmos 1429	19 Jan 1983		Grouped up to Kosmos 1436
Kosmos 1473	6 Jul 1983		Grouped up to Kosmos 1480
Kosmos 1522	5 Jan 1984		Grouped up to Kosmos 1529
Kosmos 1559	28 May 1984		Grouped up to Kosmos 1566
Kosmos 1617	15 Jan 1985		Grouped up to Kosmos 1622
Kosmos 1635	21 Mar 1985		Grouped up to Kosmos 1642
Kosmos 1690	9 Oct 1985		Grouped up to Kosmos 1695
Kosmos 1716	9 Jan 1986		Grouped up to Kosmos 1723
Kosmos 1748	6 Jun 1986		Grouped up to Kosmos 1755
Kosmos 1794	21 Nov 1986		Grouped up to Kosmos 1801
Kosmos 1827	13 Mar 1987		Grouped up to Kosmos 1832
Kosmos 1852	16 Jun 1987		Grouped up to Kosmos 1859
Kosmos 1875	7 Sep 1987		Grouped up to Kosmos 1880

KOSMOS (Military Applications)

MISSION	LAUNCH	LANDING, RECOVERY, OR RE-ENTRY	NOTES
Kosmos 31	6 Jun 1964	20 Oct 1964	
Kosmos 36	30 Jul 1964	28 Feb 1965	
Kosmos 70	2 Jul 1965	18 Dec 1966	
Kosmos 76	23 Jul 1965	16 Mar 1968	
Kosmos 93	19 Oct 1965	3 Jan 1966	
Kosmos 95	4 Nov 1965	18 Jan 1966	
Kosmos 101	21 Dec 1965	12 Jul 1966	
Kosmos 106	25 Jan 1966	14 Nov 1966	
Kosmos 116	26 Apr 1966	3 Dec 1966	
Kosmos 119	24 May 1966	30 Nov 1966	
Kosmos 123	8 Jul 1966	10 Dec 1966	
Kosmos 145	3 Mar 1967	8 Mar 1968	
Kosmos 148	16 Mar 1967	7 May 1967	
Kosmos 152	25 Mar 1967	5 Aug 1967	
Kosmos 158	15 May 1967		
Kosmos 165	12 Jun 1967	15 Jan 1968	
Kosmos 173	24 Aug 1967	17 Dec 1967	
Kosmos 176	12 Sep 1967	3 Mar 1968	
Kosmos 191	21 Nov 1967	2 Mar 1968	
Kosmos 197	26 Dec 1967	30 Jan 1968	
Kosmos 202	20 Feb 1968	24 Mar 1968	
Kosmos 204	5 Mar 1968	2 Mar 1969	
Kosmos 211	9 Apr 1968	10 Nov 1968	
Kosmos 221	24 May 1968	31 Aug 1969	
Kosmos 222	30 May 1968	11 Oct 1968	
Kosmos 233	18 Jul 1968	7 Feb 1969	
Kosmos 242	20 Sep 1968	13 Nov 1968	
Kosmos 245	3 Oct 1968	15 Jan 1969	
Kosmos 257	3 Dec 1968	5 Mar 1969	
Kosmos 265	7 Feb 1969	1 May 1969	
Kosmos 268	5 Mar 1969	9 May 1970	

TABLE 1   71

MISSION	LAUNCH	LANDING, RECOVERY, OR RE-ENTRY	NOTES
Kosmos 275	28 Mar 1969	7 Feb 1970	
Kosmos 277	4 Apr 1969	6 Jul 1969	
Kosmos 283	27 May 1969	10 Dec 1969	
Kosmos 285	3 Jun 1969	7 Oct 1969	
Kosmos 295	22 Aug 1969	1 Dec 1969	
Kosmos 303	18 Oct 1969	23 Jan 1970	
Kosmos 307	24 Oct 1969	30 Dec 1970	
Kosmos 308	4 Nov 1969	5 Jan 1970	
Kosmos 311	24 Nov 1969	10 Mar 1970	
Kosmos 314	11 Dec 1969	22 Mar 1970	
Kosmos 319	15 Jan 1970	1 Jul 1970	
Kosmos 324	27 Feb 1970	23 May 1970	
Kosmos 327	18 Mar 1970	19 Jan 1971	
Kosmos 334	23 Apr 1970	9 Aug 1970	
Kosmos 347	12 Jun 1970	7 Nov 1970	
Kosmos 351	27 Jun 1970	13 Oct 1970	
Kosmos 357	19 Aug 1970	24 Nov 1970	
Kosmos 362	16 Sep 1970	13 Oct 1971	
Kosmos 369	8 Oct 1970	22 Jan 1971	
Kosmos 372	16 Oct 1970		
Kosmos 380	24 Nov 1970	17 Jun 1971	
Kosmos 388	18 Dec 1970	10 May 1971	
Kosmos 391	14 Jan 1971	21 Feb 1972	
Kosmos 393	26 Jan 1971	16 Jun 1971	
Kosmos 407	23 Apr 1971		
Kosmos 408	24 Apr 1971	29 Dec 1971	
Kosmos 421	19 May 1971	8 Nov 1971	
Kosmos 423	27 May 1971	26 Nov 1971	
Kosmos 435	27 Aug 1971	28 Jan 1972	
Kosmos 440	24 Sep 1971	29 Oct 1972	
Kosmos 453	19 Oct 1971	19 Mar 1972	
Kosmos 455	17 Nov 1971	9 Apr 1972	
Kosmos 458	29 Nov 1971	20 Apr 1972	
Kosmos 467	17 Dec 1971	18 Apr 1972	
Kosmos 468	17 Dec 1971		
Kosmos 472	25 Jan 1972	18 Aug 1972	
Kosmos 485	11 Apr 1972	30 Aug 1972	
Kosmos 487	21 Apr 1972	24 Sep 1972	
Kosmos 494	23 Jun 1972		
Kosmos 497	30 Jun 1972	7 Nov 1973	
Kosmos 498	5 Jul 1972	25 Nov 1972	
Kosmos 501	12 Jul 1972	9 May 1974	
Kosmos 523	5 Oct 1972	7 Mar 1973	
Kosmos 524	11 Oct 1972	25 Mar 1973	
Kosmos 526	25 Oct 1972	8 Apr 1973	
Kosmos 540	25 Dec 1972		
Kosmos 545	24 Jan 1973	31 Jul 1973	
Kosmos 553	12 Apr 1973	11 Nov 1973	

MISSION	LAUNCH	LANDING, RECOVERY, OR RE-ENTRY	NOTES
Kosmos 558	17 May 1973	23 Dec 1973	
Kosmos 562	5 Jun 1973	7 Jan 1974	
Kosmos 580	22 Aug 1973	1 Apr 1974	
Kosmos 601	16 Oct 1973	15 Aug 1974	
Kosmos 608	20 Nov 1973	10 Jul 1974	
Kosmos 611	28 Nov 1973	19 Jun 1974	
Kosmos 614	4 Dec 1973		
Kosmos 615	13 Dec 1973	17 Dec 1975	
Kosmos 633	27 Feb 1974	4 Oct 1974	
Kosmos 634	5 Mar 1974	9 Oct 1974	
Kosmos 660	18 Jun 1974		
Kosmos 662	26 Jun 1974	28 Aug 1976	
Kosmos 668	25 Jul 1974	21 Feb 1975	
Kosmos 676	11 Sep 1974		
Kosmos 686	26 Sep 1974	1 May 1975	
Kosmos 687	11 Oct 1974	5 Feb 1978	
Kosmos 695	20 Nov 1974	15 Jul 1975	
Kosmos 703	21 Jan 1975	20 Nov 1975	
Kosmos 705	28 Jan 1975	18 Nov 1975	
Kosmos 725	8 Apr 1975	6 Jan 1976	
Kosmos 745	24 Jun 1975	12 Mar 1976	
Kosmos 750	17 Jul 1975	29 Sep 1977	
Kosmos 752	24 Jul 1975	28 Feb 1981	
Kosmos 773	30 Sep 1975		
Kosmos 783	28 Nov 1975		
Kosmos 801	5 Feb 1976	5 Jan 1978	
Kosmos 807	12 Mar 1976		
Kosmos 816	28 Apr 1976	24 Nov 1979	
Kosmos 818	18 May 1976	7 Mar 1977	
Kosmos 822	28 May 1976	8 Aug 1978	
Kosmos 836	29 Jun 1976		
Kosmos 841	15 Jul 1976		
Kosmos 849	18 Aug 1976	24 Aug 1978	
Kosmos 850	26 Aug 1976	16 May 1977	
Kosmos 858	29 Sep 1976		
Kosmos 885	17 Dec 1976	14 Oct 1979	
Kosmos 891	2 Feb 1977	4 Feb 1981	
Kosmos 901	5 Apr 1977	28 Jun 1978	
Kosmos 913	30 May 1977	29 Dec 1979	
Kosmos 919	18 Jun 1977	28 Aug 1978	
Kosmos 923	1 Jul 1977		
Kosmos 930	19 Jul 1977	12 May 1980	
Kosmos 933	22 Jul 1977	1 Nov 1978	
Kosmos 965	8 Dec 1977	16 Dec 1979	
Kosmos 968	16 Dec 1977		
Kosmos 990	17 Feb 1978		
Kosmos 1006	12 May 1978	14 Mar 1979	
Kosmos 1023	21 Jun 1978		

TABLE 1   73

MISSION	LAUNCH	LANDING, RECOVERY, OR RE-ENTRY	NOTES
Kosmos 1048	16 Nov 1978		
Kosmos 1065	22 Dec 1978	1 Aug 1979	
Kosmos 1075	8 Feb 1979	19 Oct 1981	
Kosmos 1110	28 Jun 1979		
Kosmos 1112	6 Jul 1979	21 Jan 1980	
Kosmos 1125	28 Aug 1979		
Kosmos 1140	11 Oct 1979		
Kosmos 1146	5 Dec 1979	25 Nov 1981	
Kosmos 1169	27 Mar 1980	3 Mar 1983	
Kosmos 1179	14 May 1980	18 Jul 1989	
Kosmos 1186	6 Jun 1980	1 Jan 1982	
Kosmos 1190	1 Jul 1980		
Kosmos 1204	31 Jul 1980	23 Feb 1981	
Kosmos 1215	14 Oct 1980	12 May 1983	
Kosmos 1238	16 Jan 1981		
Kosmos 1263	9 Apr 1981		
Kosmos 1269	7 May 1981		
Kosmos 1302	28 Aug 1981		
Kosmos 1310	23 Sep 1981		
Kosmos 1311	28 Sep 1981	28 Aug 1983	
Kosmos 1331	7 Jan 1982		
Kosmos 1335	29 Jan 1982	5 Apr 1987	
Kosmos 1354	28 Apr 1982		
Kosmos 1371	1 Jun 1982		
Kosmos 1410	24 Sep 1982		
Kosmos 1420	11 Nov 1982		
Kosmos 1452	12 Apr 1983		
Kosmos 1463	19 May 1983		
Kosmos 1486	3 Aug 1983		
Kosmos 1503	12 Oct 1983		
Kosmos 1508	11 Nov 1983		
Kosmos 1538	21 Feb 1984		
Kosmos 1570	8 Jun 1984		
Kosmos 1578	28 Jun 1984		
Kosmos 1615	20 Dec 1984		
Kosmos 1624	17 Jan 1985		
Kosmos 1631	27 Feb 1985		
Kosmos 1680	4 Sep 1985		
Kosmos 1741	18 Apr 1986		
Kosmos 1763	16 Jul 1986		
Kosmos 1776	3 Sep 1986		
Kosmos 1777	10 Sep 1986		
Kosmos 1788	27 Oct 1986		
Kosmos 1814	21 Jan 1987		
Kosmos 1815	22 Jan 1987		
Kosmos 1850	9 Jun 1987		
Kosmos 1868	14 Jul 1987		
Kosmos 1898	1 Dec 1987		

MISSION	LAUNCH	LANDING, RECOVERY, OR RE-ENTRY	NOTES
KOSMOS (Navigation)			
Kosmos 192	23 Nov 1967		
Kosmos 220	7 May 1968		
Kosmos 292	14 Aug 1969		
Kosmos 304	21 Oct 1969		
Kosmos 332	11 Apr 1970		
Kosmos 371	12 Oct 1970		
Kosmos 385	12 Dec 1970		
Kosmos 422	22 May 1971		
Kosmos 465	15 Dec 1971		
Kosmos 475	25 Feb 1972		
Kosmos 489	6 May 1972		
Kosmos 514	16 Aug 1972		
Kosmos 574	20 Jun 1973		
Kosmos 586	14 Sep 1973		
Kosmos 627	29 Dec 1973		
Kosmos 628	17 Jan 1974		
Kosmos 663	27 Jun 1974		
Kosmos 689	18 Oct 1974		
Kosmos 700	26 Dec 1974		
Kosmos 726	11 Apr 1975		
Kosmos 729	22 Apr 1975		
Kosmos 755	14 Aug 1975		
Kosmos 778	4 Nov 1975		
Kosmos 789	20 Jan 1976		
Kosmos 800	3 Feb 1976		
Kosmos 823	2 Jun 1976		
Kosmos 846	29 Jul 1976		
Kosmos 864	29 Oct 1976		
Kosmos 883	15 Dec 1976		
Kosmos 887	28 Dec 1976		
Kosmos 890	20 Jan 1977		
Kosmos 894	21 Feb 1977		
Kosmos 926	8 Jul 1977		
Kosmos 928	13 Jul 1977		
Kosmos 951	13 Sep 1977		
Kosmos 962	28 Oct 1977		
Kosmos 971	23 Dec 1977		
Kosmos 985	17 Jan 1978		
Kosmos 991	28 Feb 1978		
Kosmos 994	15 Mar 1978		
Kosmos 996	28 Mar 1978		
Kosmos 1000	31 Mar 1978		
Kosmos 1011	23 May 1978		
Kosmos 1027	27 Jul 1978		
Kosmos 1064	20 Dec 1978		
Kosmos 1089	21 Mar 1979		
Kosmos 1091	7 Apr 1979		

TABLE 1   75

MISSION	LAUNCH	LANDING, RECOVERY, OR RE-ENTRY	NOTES
Kosmos 1092	11 Apr 1979		
Kosmos 1104	31 May 1979		
Kosmos 1141	16 Oct 1979		
Kosmos 1150	14 Jan 1980		
Kosmos 1153	25 Jan 1980		
Kosmos 1168	17 Mar 1980		
Kosmos 1181	20 May 1980		
Kosmos 1225	5 Dec 1980		
Kosmos 1226	10 Dec 1980		
Kosmos 1244	12 Feb 1981		
Kosmos 1275	14 Jun 1981		
Kosmos 1295	12 Aug 1981		
Kosmos 1304	4 Sep 1981		
Kosmos 1308	18 Sep 1981		
Kosmos 1333	14 Jan 1982		
Kosmos 1339	17 Feb 1982		
Kosmos 1344	24 Mar 1982		
Kosmos 1349	8 Apr 1982		
Kosmos 1383	29 Jun 1982		
Kosmos 1386	7 Jul 1982		
Kosmos 1413	12 Oct 1982		
Kosmos 1414	12 Oct 1982		
Kosmos 1415	12 Oct 1982		
Kosmos 1417	19 Oct 1982		
Kosmos 1428	12 Jan 1983		
Kosmos 1447	24 Mar 1983		
Kosmos 1448	30 Mar 1983		
Kosmos 1459	6 May 1983		
Kosmos 1464	24 May 1983		
Kosmos 1490	10 Aug 1983		
Kosmos 1491	10 Aug 1983		
Kosmos 1492	10 Aug 1983		
Kosmos 1506	26 Oct 1983		
Kosmos 1513	8 Dec 1983		
Kosmos 1519	29 Dec 1983		
Kosmos 1520	29 Dec 1983		
Kosmos 1521	29 Dec 1983		
Kosmos 1531	11 Jan 1984		
Kosmos 1535	2 Feb 1984		
Kosmos 1550	11 May 1984		
Kosmos 1553	17 May 1984		
Kosmos 1554	19 May 1984		
Kosmos 1555	19 May 1984		
Kosmos 1556	19 May 1984		
Kosmos 1574	21 Jun 1984		
Kosmos 1577	27 Jun 1984		
Kosmos 1593	4 Sep 1984		
Kosmos 1594	4 Sep 1984		

MISSION	LAUNCH	LANDING, RECOVERY, OR RE-ENTRY	NOTES
Kosmos 1595	4 Sep 1984		
Kosmos 1598	13 Sep 1984		
Kosmos 1605	11 Oct 1984		
Kosmos 1610	15 Nov 1984		
Kosmos 1627	1 Feb 1985		
Kosmos 1634	14 Mar 1985		
Kosmos 1650	17 May 1985		
Kosmos 1651	17 May 1985		
Kosmos 1652	17 May 1985		
Kosmos 1655	30 May 1985		
Kosmos 1666	8 Jul 1985		
Kosmos 1670	1 Aug 1985		
Kosmos 1704	28 Nov 1985		
Kosmos 1709	19 Dec 1985		
Kosmos 1710	24 Dec 1985		
Kosmos 1711	24 Dec 1985		
Kosmos 1712	24 Dec 1985		
Kosmos 1725	17 Jan 1986		
Kosmos 1727	23 Jan 1986		
Kosmos 1745	23 May 1986		
Kosmos 1759	18 Jun 1986		
Kosmos 1778	16 Sep 1986		
Kosmos 1779	16 Sep 1986		
Kosmos 1780	16 Sep 1986		
Kosmos 1791	13 Nov 1986		
Kosmos 1802	24 Nov 1986		
Kosmos 1808	17 Dec 1986		
Kosmos 1816	29 Jan 1987		
Kosmos 1821	18 Feb 1987		
Kosmos 1861	23 Jun 1987		
Kosmos 1864	6 Jul 1987		
Kosmos 1883	16 Sep 1987		
Kosmos 1884	16 Sep 1987		
Kosmos 1885	16 Sep 1987		
Kosmos 1891	14 Oct 1987		
Kosmos 1904	23 Dec 1987		
KOSMOS (Ocean Surveillance)			
Kosmos 198	28 Dec 1967		
Kosmos 209	22 Mar 1968		
Kosmos 367	3 Oct 1970		
Kosmos 402	1 Apr 1971		
Kosmos 469	25 Dec 1971		
Kosmos 516	21 Aug 1972		
Kosmos 626	27 Dec 1973		
Kosmos 651	15 May 1974		
Kosmos 654	17 May 1974		
Kosmos 699	24 Dec 1974	16 Oct 1977	
Kosmos 723	2 Apr 1975		

TABLE 1   77

MISSION	LAUNCH	LANDING, RECOVERY, OR RE-ENTRY	NOTES
Kosmos 724	7 Apr 1975		
Kosmos 777	29 Oct 1975	3 Jun 1976	
Kosmos 785	12 Dec 1975		
Kosmos 838	2 Jul 1976	23 Aug 1977	
Kosmos 860	17 Oct 1976		
Kosmos 861	21 Oct 1976		
Kosmos 868	26 Nov 1976	8 Jul 1978	
Kosmos 937	24 Aug 1977	19 Oct 1978	
Kosmos 952	16 Sep 1977		
Kosmos 954	18 Sep 1977	24 Jan 1978	
Kosmos 1025	28 Jun 1978		
Kosmos 1045	26 Oct 1978		
Kosmos 1094	18 Apr 1979	7 Nov 1979	
Kosmos 1096	25 Apr 1979	24 Nov 1979	
Kosmos 1167	14 Mar 1980	1 Oct 1981	
Kosmos 1176	29 Apr 1980		
Kosmos 1220	4 Nov 1980		
Kosmos 1249	5 Mar 1981		
Kosmos 1260	20 Mar 1981	22 May 1982	
Kosmos 1266	21 Apr 1981		
Kosmos 1286	4 Aug 1981	16 Oct 1982	
Kosmos 1299	24 Aug 1981		
Kosmos 1306	4 Sep 1981	16 Jul 1982	
Kosmos 1337	11 Feb 1982	25 Jul 1982	
Kosmos 1355	29 Apr 1982	27 Aug 1983	
Kosmos 1365	14 May 1982		
Kosmos 1372	1 Jun 1982		
Kosmos 1402	30 Aug 1982	23 Jan 1983	
Kosmos 1405	4 Sep 1982	5 Feb 1984	
Kosmos 1412	2 Oct 1982		
Kosmos 1461	7 May 1983		
Kosmos 1507	29 Oct 1983	19 Aug 1987	
Kosmos 1567	30 May 1984		
Kosmos 1579	29 Jun 1984		
Kosmos 1588	7 Aug 1984		
Kosmos 1607	31 Oct 1984		
Kosmos 1646	18 Apr 1985		
Kosmos 1677	23 Aug 1985		
Kosmos 1682	19 Sep 1985		
Kosmos 1735	27 Feb 1986		
Kosmos 1736	21 Mar 1986		
Kosmos 1737	25 Mar 1986	3 Dec 1986	
Kosmos 1769	4 Aug 1986	21 Sep 1987	
Kosmos 1771	20 Aug 1986		
Kosmos 1818	1 Feb 1987		
Kosmos 1834	8 Apr 1987		
Kosmos 1860	18 Jun 1987		
Kosmos 1867	10 Jul 1987		

MISSION	LAUNCH	LANDING, RECOVERY, OR RE-ENTRY	NOTES
Kosmos 1890	10 Oct 1987		
Kosmos 1900	12 Dec 1987		

KOSMOS (Science, Technology)

MISSION	LAUNCH	LANDING, RECOVERY, OR RE-ENTRY	NOTES
Kosmos 1	16 Mar 1962	25 May 1962	
Kosmos 2	6 Apr 1962	19 Aug 1963	
Kosmos 3	24 Apr 1962	17 Oct 1962	
Kosmos 5	28 May 1962	2 May 1963	
Kosmos 6	30 Jun 1962	8 Aug 1962	
Kosmos 8	18 Aug 1962	17 Aug 1963	
Kosmos 11	20 Oct 1962	18 May 1964	
Kosmos 14	13 Apr 1963	29 Aug 1963	
Kosmos 17	22 May 1963	2 Jun 1965	
Kosmos 19	6 Aug 1963	30 Mar 1964	
Kosmos 23	13 Dec 1963	27 Mar 1964	
Kosmos 25	27 Feb 1964	21 Nov 1964	
Kosmos 26	18 Mar 1964	28 Sep 1964	
Kosmos 41	22 Aug 1964		
Kosmos 43	22 Aug 1964	27 Dec 1965	
Kosmos 44	28 Aug 1964		
Kosmos 49	24 Oct 1964	21 Aug 1965	
Kosmos 51	10 Dec 1964	14 Nov 1965	
Kosmos 53	30 Jan 1965	12 Aug 1966	
Kosmos 58	26 Feb 1965		
Kosmos 97	26 Nov 1965	2 Apr 1967	
Kosmos 100	17 Dec 1965		
Kosmos 102	28 Dec 1965	13 Jan 1966	
Kosmos 108	11 Feb 1966	21 Nov 1966	
Kosmos 110	22 Feb 1966	16 Mar 1966	
Kosmos 118	11 May 1966		
Kosmos 122	25 Jun 1966		
Kosmos 125	20 Jul 1966	2 Aug 1966	
Kosmos 135	12 Dec 1966	12 Apr 1967	
Kosmos 137	21 Dec 1966	23 Nov 1967	
Kosmos 142	14 Feb 1967	6 Jul 1967	
Kosmos 144	28 Feb 1967	14 Sep 1982	
Kosmos 149	21 Mar 1967	7 Apr 1967	
Kosmos 156	27 Apr 1967		
Kosmos 163	5 Jun 1967	11 Oct 1967	
Kosmos 166	16 Jun 1967	25 Oct 1967	
Kosmos 184	24 Oct 1967		
Kosmos 196	19 Dec 1967	7 Jul 1968	
Kosmos 203	20 Feb 1968		
Kosmos 206	14 Mar 1968		
Kosmos 215	19 Apr 1968	30 Jun 1968	
Kosmos 219	26 Apr 1968	2 Mar 1969	
Kosmos 225	11 Jun 1968	2 Nov 1968	
Kosmos 226	12 Jun 1968	18 Oct 1983	
Kosmos 230	5 Jul 1968	2 Nov 1968	

TABLE 1   79

MISSION	LAUNCH	LANDING, RECOVERY, OR RE-ENTRY	NOTES
Kosmos 256	30 Nov 1968		
Kosmos 259	14 Dec 1968	5 May 1969	
Kosmos 261	20 Dec 1968	12 Feb 1969	
Kosmos 262	26 Dec 1968	18 Jul 1969	
Kosmos 272	17 Mar 1969		
Kosmos 312	24 Nov 1969		
Kosmos 320	16 Jan 1970	10 Feb 1970	
Kosmos 321	20 Jan 1970	23 Mar 1970	
Kosmos 335	24 Apr 1970	22 Jun 1970	
Kosmos 348	13 Jun 1970	25 Jul 1970	
Kosmos 356	10 Aug 1970	2 Oct 1970	
Kosmos 378	17 Nov 1970	17 Aug 1972	
Kosmos 381	2 Dec 1970		
Kosmos 409	28 Apr 1971		
Kosmos 426	4 Jun 1971		
Kosmos 457	20 Nov 1971		
Kosmos 461	2 Dec 1971	21 Feb 1979	
Kosmos 480	25 Mar 1972		
Kosmos 481	25 Mar 1972	2 Sep 1972	
Kosmos 539	21 Dec 1972		
Kosmos 546	26 Jan 1973		
Kosmos 585	8 Sep 1973		
Kosmos 605	31 Oct 1973	22 Nov 1973	
Kosmos 637	26 Mar 1974		
Kosmos 650	29 Apr 1974		
Kosmos 675	29 Aug 1974		
Kosmos 690	22 Oct 1974	12 Nov 1974	
Kosmos 708	12 Feb 1975		
Kosmos 770	24 Sep 1975		
Kosmos 782	25 Nov 1975	15 Dec 1975	
Kosmos 842	21 Jul 1976		
Kosmos 893	15 Feb 1977	6 Oct 1984	
Kosmos 900	29 Mar 1977	11 Oct 1979	
Kosmos 906	27 Apr 1977	23 Mar 1980	
Kosmos 911	25 May 1977		
Kosmos 921	24 Jun 1977		
Kosmos 936	3 Aug 1977	22 Aug 1977	
Kosmos 956	24 Sep 1977	27 Jun 1982	
Kosmos 963	24 Nov 1977		
Kosmos 972	27 Dec 1977		
Kosmos 1067	26 Dec 1978		
Kosmos 1076	12 Feb 1979		
Kosmos 1129	25 Sep 1979	14 Oct 1979	
Kosmos 1151	23 Jan 1980		
Kosmos 1273	22 May 1981	4 Jun 1981	
Kosmos 1276	16 Jun 1981	28 Jun 1981	
Kosmos 1280	2 Jul 1981	15 Jul 1981	
Kosmos 1283	17 Jul 1981	31 Jul 1981	

MISSION	LAUNCH	LANDING, RECOVERY, OR RE-ENTRY	NOTES
Kosmos 1284	29 Jul 1981	12 Aug 1981	
Kosmos 1300	24 Aug 1981		
Kosmos 1301	27 Aug 1981	10 Sep 1981	
Kosmos 1312	30 Sep 1981		
Kosmos 1314	9 Oct 1981	22 Oct 1981	
Kosmos 1328	3 Dec 1981		
Kosmos 1378	10 Jun 1982		
Kosmos 1408	16 Sep 1982		
Kosmos 1484	24 Jul 1983		
Kosmos 1500	28 Sep 1983		
Kosmos 1510	24 Nov 1983		
Kosmos 1514	14 Dec 1983	19 Dec 1983	
Kosmos 1557	22 May 1984	4 Jun 1984	
Kosmos 1572	15 Jun 1984	29 Jun 1984	
Kosmos 1575	22 Jun 1984	7 Jul 1984	
Kosmos 1584	27 Jul 1984	10 Aug 1984	
Kosmos 1589	8 Aug 1984		
Kosmos 1590	16 Aug 1984	30 Aug 1984	
Kosmos 1591	30 Aug 1984	13 Sep 1984	
Kosmos 1597	13 Sep 1984	26 Sep 1984	
Kosmos 1601	27 Sep 1984		
Kosmos 1602	28 Sep 1984		
Kosmos 1603	28 Sep 1984		
Kosmos 1653	22 May 1985	5 Jun 1985	
Kosmos 1657	7 Jun 1985	21 Jun 1985	
Kosmos 1660	14 Jun 1985		
Kosmos 1662	19 Jun 1985		
Kosmos 1663	21 Jun 1985	5 Jul 1985	
Kosmos 1667	10 Jul 1985	17 Jul 1985	
Kosmos 1672	7 Aug 1985	21 Aug 1985	
Kosmos 1678	29 Aug 1985	12 Sep 1985	
Kosmos 1681	6 Sep 1985	19 Sep 1985	
Kosmos 1688	2 Oct 1985		
Kosmos 1689	3 Oct 1985		
Kosmos 1708	13 Dec 1985	27 Dec 1985	
Kosmos 1732	11 Feb 1986		
Kosmos 1762	10 Jul 1986	24 Jul 1986	
Kosmos 1766	28 Jul 1986		
Kosmos 1786	22 Oct 1986		
Kosmos 1803	2 Dec 1986		
Kosmos 1809	18 Dec 1986		
Kosmos 1820	14 Feb 1987	6 Mar 1987	
Kosmos 1823	20 Feb 1987		
Kosmos 1841	24 Apr 1987	8 May 1987	
Kosmos 1869	14 Jul 1987		
Kosmos 1870	25 Jul 1987	29 Jul 1989	
Kosmos 1871	1 Aug 1987	10 Aug 1987	
Kosmos 1873	28 Aug 1987	14 Sep 1987	

TABLE 1   81

MISSION	LAUNCH	LANDING, RECOVERY, OR RE-ENTRY	NOTES
Kosmos 1882	15 Sep 1987	6 Oct 1987	
Kosmos 1887	29 Sep 1987	12 Oct 1987	
Kosmos 1906	26 Dec 1987		

KOSMOS (Tests for Various Space Vehicles)

MISSION	LAUNCH	LANDING, RECOVERY, OR RE-ENTRY	NOTES
Kosmos 47	6 Oct 1964	7 Oct 1964	Voskhod test
Kosmos 57	22 Feb 1965	22 Feb 1965	Voskhod test
Kosmos 133	28 Nov 1966	30 Nov 1966	Soyuz test
Kosmos 140	7 Feb 1967	9 Feb 1967	Soyuz test
Kosmos 186	27 Oct 1967	31 Oct 1967	Soyuz test
Kosmos 188	30 Oct 1967	2 Nov 1967	Soyuz test
Kosmos 212	14 Apr 1968	19 Apr 1968	Soyuz test
Kosmos 213	15 Apr 1968	20 Apr 1968	Soyuz test
Kosmos 238	28 Aug 1968	1 Sep 1968	Soyuz test
Kosmos 379	24 Nov 1970	21 Sep 1983	Test, manned lunar probe
Kosmos 382	2 Dec 1970		Test, manned lunar probe
Kosmos 398	26 Feb 1971		Test, manned lunar probe
Kosmos 434	12 Aug 1971	12 Aug 1971	Test, manned lunar probe
Kosmos 496	26 Jun 1972	2 Jul 1972	Soyuz test
Kosmos 573	15 Jun 1973	17 Jun 1973	Soyuz test
Kosmos 613	30 Nov 1973	29 Jan 1974	Soyuz test
Kosmos 638	3 Apr 1974	13 Apr 1974	Soyuz test
Kosmos 656	27 May 1974	29 May 1974	Soyuz test
Kosmos 670	6 Aug 1974	9 Aug 1974	Soyuz test
Kosmos 672	12 Aug 1974	18 Aug 1974	Soyuz test
Kosmos 772	29 Sep 1975	2 Oct 1975	Soyuz test
Kosmos 869	29 Nov 1976	17 Dec 1976	Soyuz test
Kosmos 881	15 Dec 1976	15 Dec 1976	Test, reusable spacecraft
Kosmos 882	15 Dec 1976	15 Dec 1976	Test, reusable spacecraft
Kosmos 929	17 Jul 1977	2 Feb 1978	Modulny space station test
Kosmos 997	30 Mar 1978	30 Mar 1978	Test, reusable spacecraft
Kosmos 998	30 Mar 1978	30 Mar 1978	Test, reusable spacecraft
Kosmos 1001	4 Apr 1978	15 Apr 1978	Soyuz T test
Kosmos 1074	31 Jan 1979	1 Apr 1979	Soyuz T test
Kosmos 1100	22 May 1979	23 May 1979	Test, reusable spacecraft
Kosmos 1101	22 May 1979	23 May 1979	Test, reusable spacecraft
Kosmos 1267	25 Apr 1981	29 Jul 1982	Modulny space station test
Kosmos 1374	3 Jun 1982	3 Jun 1982	Test, reusable spacecraft
Kosmos 1443	2 Mar 1983	19 Sep 1983	Modulny space station test
Kosmos 1445	15 Mar 1983	15 Mar 1983	Test, reusable spacecraft
Kosmos 1517	27 Dec 1983	27 Dec 1983	Test, reusable spacecraft
Kosmos 1614	19 Dec 1984	19 Dec 1984	Test, reusable spacecraft
Kosmos 1669	19 Jul 1985	30 Aug 1985	Soyuz TM test
Kosmos 1686	27 Sep 1985		Modulny space station test

METEOR (Science, Technology)

MISSION	LAUNCH	LANDING, RECOVERY, OR RE-ENTRY	NOTES
Meteor 1-1	26 Mar 1969		
Meteor 1-2	6 Oct 1969		
Meteor 1-3	17 Mar 1970	17 Nov 1983	

MISSION	LAUNCH	LANDING, RECOVERY, OR RE-ENTRY	NOTES
Meteor 1-4	28 Apr 1970		
Meteor 1-5	23 Jun 1970		
Meteor 1-6	15 Oct 1970		
Meteor 1-7	20 Jan 1971		
Meteor 1-8	17 Apr 1971		
Meteor 1-9	16 Jul 1971		
Meteor 1-10	29 Dec 1971		
Meteor 1-11	30 Mar 1972		
Meteor 1-12	30 Jun 1972		
Meteor 1-13	26 Oct 1972		
Meteor 1-14	20 Mar 1973		
Meteor 1-15	29 May 1973		
Meteor 1-16	5 Mar 1974		
Meteor 1-17	24 Apr 1974		
Meteor 1-18	9 Jul 1974		
Meteor 1-19	28 Oct 1974		
Meteor 1-20	17 Dec 1974	8 Dec 1980	
Meteor 1-21	1 Apr 1975		
Meteor 2-1	11 Jul 1975		
Meteor 1-22	18 Sep 1975		
Meteor 1-23	25 Dec 1975		
Meteor 1-24	7 Apr 1976		
Meteor 1-25	15 May 1976		
Meteor 1-26	15 Oct 1976		
Meteor 2-2	6 Jan 1977		
Meteor 1-27	5 Apr 1977		
Meteor 1-28	29 Jun 1977		
Meteor 2-3	14 Dec 1977		
Meteor 1-29	25 Jan 1979		
Meteor 2-4	1 Mar 1979		
Meteor 2-5	31 Oct 1979		
Meteor 1-30	18 Jun 1980		
Meteor 2-6	9 Sep 1980		
Meteor 2-7	14 May 1981		
Meteor 1-31	10 Jul 1981		
Meteor 2-8	25 Mar 1982		
Meteor 2-9	14 Dec 1982		
Meteor 2-10	28 Oct 1983		
Meteor 2-11	5 Jul 1984		
Meteor 2-12	6 Feb 1985		
Meteor 3-1	23 Oct 1985		
Meteor 2-13	26 Dec 1985		
Meteor 2-14	27 May 1986		
Meteor 2-15	5 Jan 1987		
Meteor 2-16	18 Aug 1987		

MOLNIYA (Communications)

Molniya 1-1	23 Apr 1965	16 Aug 1979	
Molniya 1-2	14 Oct 1965	17 Mar 1967	

TABLE 1 83

MISSION	LAUNCH	LANDING, RECOVERY, OR RE-ENTRY	NOTES
Molniya 1-3	25 Apr 1966	11 Jun 1973	
Molniya 1-4	20 Oct 1966	11 Sep 1968	
Molniya 1-5	24 May 1967	26 Nov 1971	
Molniya 1-6	3 Oct 1967	4 Mar 1969	
Molniya 1-7	22 Oct 1967	31 Dec 1969	
Molniya 1-8	21 Apr 1968	29 Jan 1974	
Molniya 1-9	5 Jul 1968	15 May 1971	
Molniya 1-10	5 Oct 1968	16 Jul 1976	
Molniya 1-11	11 Apr 1969	17 Apr 1974	
Molniya 1-12	22 Jul 1969	18 Jun 1971	
Molniya 1-13	19 Feb 1970	29 Sep 1975	
Molniya 1-14	26 Jun 1970	16 Feb 1976	
Molniya 1-15	29 Sep 1970	20 Mar 1976	
Molniya 1-16	27 Nov 1970	25 Nov 1975	
Molniya 1-17	25 Dec 1970	22 Dec 1975	
Molniya 1-18	28 Jul 1971	19 Jul 1977	
Molniya 2-1	24 Nov 1971	10 May 1976	
Molniya 1-19	19 Dec 1971	13 Apr 1977	
Molniya 1-20	4 Apr 1972	30 Jan 1974	
Molniya 2-2	19 May 1972	22 Mar 1977	
Molniya 2-3	30 Sep 1972	12 Jan 1978	
Molniya 1-21	14 Oct 1972	1 Nov 1977	
Molniya 1-22	2 Dec 1972	11 Feb 1976	
Molniya 2-4	12 Dec 1972	22 Jan 1975	
Molniya 1-23	3 Feb 1973	23 Oct 1977	
Molniya 2-5	5 Apr 1973	6 Jan 1979	
Molniya 2-6	11 Jul 1973	5 Aug 1978	
Molniya 1-24	30 Aug 1973	5 Dec 1979	
Molniya 2-7	19 Oct 1973	8 Jul 1983	
Molniya 1-25	14 Nov 1973	26 May 1979	
Molniya 1-26	30 Nov 1973	9 Jun 1985	
Molniya 2-8	25 Dec 1973	24 Nov 1984	
Molniya 1-27	20 Apr 1974	17 Nov 1983	
Molniya 2-9	26 Apr 1974		
Molniya 2-10	23 Jul 1974		
Molniya 1S	29 Jul 1974		
Molniya 1-28	24 Oct 1974	29 Dec 1985	
Molniya 3-1	21 Nov 1974	15 May 1986	
Molniya 2-11	21 Dec 1974		
Molniya 2-12	6 Feb 1975	4 Jul 1985	
Molniya 3-2	14 Apr 1975		
Molniya 1-29	29 Apr 1975		
Molniya 1-30	5 Jun 1975	12 Aug 1987	
Molniya 2-13	8 Jul 1975		
Molniya 1-31	2 Sep 1975	19 Nov 1985	
Molniya 2-14	9 Sep 1975		
Molniya 3-3	14 Nov 1975		
Molniya 2-15	17 Dec 1975	7 Mar 1987	

MISSION	LAUNCH	LANDING, RECOVERY, OR RE-ENTRY	NOTES
Molniya 3-4	27 Dec 1975	12 Aug 1986	
Molniya 1-32	22 Jan 1976		
Molniya 1-33	11 Mar 1976		
Molniya 1-34	19 Mar 1976	14 May 1985	
Molniya 3-5	12 May 1976		
Molniya 1-35	23 Jul 1976	29 May 1987	
Molniya 2-16	2 Dec 1976		
Molniya 3-6	28 Dec 1976		
Molniya 2-17	11 Feb 1977		
Molniya 1-36	24 Mar 1977		
Molniya 3-7	28 Apr 1977		
Molniya 1-37	24 Jun 1977		
Molniya 1-38	30 Aug 1977		
Molniya 3-8	28 Oct 1977		
Molniya 3-9	24 Jan 1978		
Molniya 1-39	3 Mar 1978		
Molniya 1-40	2 Jun 1978		
Molniya 1-41	14 Jul 1978		
Molniya 1-42	22 Aug 1978		
Molniya 3-10	13 Oct 1978		
Molniya 3-11	18 Jan 1979		
Molniya 1-43	12 Apr 1979		
Molniya 3-12	5 Jun 1979		
Molniya 1-44	31 Jul 1979		
Molniya 1-45	20 Oct 1979		
Molniya 1-46	11 Jan 1980		
Molniya 1-47	21 Jun 1980		
Molniya 3-13	18 Jul 1980		
Molniya 1-48	16 Nov 1980		
Molniya 3-14	9 Jan 1981		
Molniya 1-49	30 Jan 1981		
Molniya 3-15	24 Mar 1981		
Molniya 3-16	9 Jun 1981		
Molniya 1-50	24 Jun 1981		
Molniya 3-17	17 Oct 1981		
Molniya 1-51	17 Nov 1981		
Molniya 1-52	23 Dec 1981		
Molniya 1-53	26 Feb 1982		
Molniya 3-18	24 Mar 1982		
Molniya 1-54	28 May 1982		
Molniya 1-55	21 Jul 1982		
Molniya 3-19	27 Aug 1982		
Molniya 3-20	11 Mar 1983		
Molniya 1-56	16 Mar 1983		
Molniya 1-57	2 Apr 1983		
Molniya 1-58	19 Jul 1983		
Molniya 3-21	30 Aug 1983		
Molniya 1-59	23 Nov 1983		

TABLE 1   85

MISSION	LAUNCH	LANDING, RECOVERY, OR RE-ENTRY	NOTES
Molniya 3-22	21 Dec 1983		
Molniya 1-60	16 Mar 1984		
Molniya 1-61	10 Aug 1984		
Molniya 1-62	24 Aug 1984		
Molniya 1-63	14 Dec 1984		
Molniya 3-23	16 Jan 1985		
Molniya 3-24	29 May 1985		
Molniya 3-25	17 Jul 1985		
Molniya 1-64	22 Aug 1985		
Molniya 3-26	3 Oct 1985		
Molniya 1-65	23 Oct 1985		
Molniya 1-66	28 Oct 1985		
Molniya 3-27	24 Dec 1985		
Molniya 3-28	18 Apr 1986		
Molniya 3-29	19 Jun 1986		
Molniya 1-67	30 Jul 1986		
Molniya 1-68	5 Sep 1986		
Molniya 3-30	20 Oct 1986		
Molniya 1-69	15 Nov 1986		
Molniya 1-70	26 Dec 1986		
Molniya 3-31	22 Jan 1987		

OTHER SATELLITES

MISSION	LAUNCH	LANDING, RECOVERY, OR RE-ENTRY	NOTES
Unnamed	17 Sep 1966	11 Nov 1966	Possibly a FOBS-Kosmos
Unnamed	2 Nov 1966	29 Nov 1966	Possibly a FOBS-Kosmos
Astron	23 Mar 1983		Astronomy

POLYOT (Science, Technology)

MISSION	LAUNCH	LANDING, RECOVERY, OR RE-ENTRY	NOTES
Polyot 1	1 Nov 1963	16 Oct 1982	
Polyot 2	12 Apr 1964	8 Jun 1966	

PROGNOZ (Science, Technology)

MISSION	LAUNCH	LANDING, RECOVERY, OR RE-ENTRY	NOTES
Prognoz 1	14 Apr 1972		
Prognoz 2	29 Jun 1972	15 Dec 1982	
Prognoz 3	15 Feb 1973		
Prognoz 4	22 Dec 1975		
Prognoz 5	25 Nov 1976		
Prognoz 6	22 Sep 1977		
Prognoz 7	30 Oct 1978	22 Oct 1980	
Prognoz 8	25 Dec 1980		
Prognoz 9	1 Jul 1983	19 Jul 1983	
Prognoz 10	26 Apr 1988		

PROGRESS (Supply Vehicles for Manned Space Stations)

MISSION	LAUNCH	LANDING, RECOVERY, OR RE-ENTRY	NOTES
Progress 1	20 Jan 1978	8 Feb 1978	
Progress 2	7 Jul 1978	4 Aug 1978	
Progress 3	7 Aug 1978	23 Aug 1978	
Progress 4	4 Oct 1978	26 Oct 1978	
Progress 5	12 Mar 1979	5 Apr 1979	
Progress 6	12 May 1979	9 Jun 1979	

MISSION	LAUNCH	LANDING, RECOVERY, OR RE-ENTRY	NOTES
Progress 7	27 Jun 1979	20 Jul 1979	
Progress 8	27 Mar 1980	26 Apr 1980	
Progress 9	27 Apr 1980	22 May 1980	
Progress 10	29 Jun 1980	19 Jul 1980	
Progress 11	28 Sep 1980	11 Dec 1980	
Progress 12	24 Jan 1981	20 Mar 1981	
Progress 13	23 May 1982	6 Jun 1982	
Progress 14	10 Jul 1982	13 Aug 1982	
Progress 15	18 Sep 1982	16 Oct 1982	
Progress 16	31 Oct 1982	14 Dec 1982	
Progress 17	17 Aug 1983	18 Sep 1983	
Progress 18	20 Oct 1983	16 Nov 1983	
Progress 19	21 Feb 1984	1 Apr 1984	
Progress 20	15 Apr 1984	7 May 1984	
Progress 21	7 May 1984	26 May 1984	
Progress 22	28 May 1984	15 Jul 1984	
Progress 23	14 Aug 1984	28 Aug 1984	
Progress 24	21 Jun 1985	15 Jul 1985	
Progress 25	19 Mar 1986	21 Apr 1986	
Progress 26	23 Apr 1986	23 Jun 1986	
Progress 27	16 Jan 1987	25 Feb 1987	
Progress 28	3 Mar 1987	28 Mar 1987	
Progress 29	21 Apr 1987	11 May 1987	
Progress 30	19 May 1987	19 Jul 1987	
Progress 31	3 Aug 1987	23 Sep 1987	
Progress 32	23 Sep 1987	19 Nov 1987	
Progress 33	20 Nov 1987	19 Dec 1987	

PROTON (Science, Technology)

Proton 1	16 Jul 1965	11 Oct 1965	
Proton 2	2 Nov 1965	6 Feb 1966	
Proton 3	6 Jul 1966	16 Sep 1966	
Proton 4	16 Nov 1968	24 Jul 1969	

RADIO (Communications)

Radio 1	26 Oct 1978		
Radio 2	26 Oct 1978		
Radio 3	17 Dec 1981		
Radio 4	17 Dec 1981		
Radio 5	17 Dec 1981		
Radio 6	17 Dec 1981		
Radio 7	17 Dec 1981		
Radio 8	17 Dec 1981		

RADUGA (Communications)

Raduga 1	22 Dec 1975		
Raduga 2	11 Sep 1976		
Raduga 3	23 Jul 1977		
Raduga 4	18 Jul 1978		
Raduga 5	25 Apr 1979		

TABLE 1   87

MISSION	LAUNCH	LANDING, RECOVERY, OR RE-ENTRY	NOTES
Raduga 6	20 Feb 1980		
Raduga 7	5 Oct 1980		
Raduga 8	18 Mar 1981		
Raduga 9	30 Jul 1981		
Raduga 10	9 Oct 1981		
Raduga 11	26 Nov 1982		
Raduga 12	8 Apr 1983		
Raduga 13	25 Aug 1983		
Raduga 14	15 Feb 1984		
Raduga 15	22 Jun 1984		
Raduga 16	8 Aug 1985		
Raduga 17	15 Nov 1985		
Raduga 18	17 Jan 1986		
Raduga 19	25 Oct 1986		
Raduga 20	19 Mar 1987		
Raduga 21	10 Dec 1987		

SOYUZ (Precursors to Manned Flights)

MISSION	LAUNCH	LANDING, RECOVERY, OR RE-ENTRY	NOTES
Soyuz 2	25 Oct 1968	28 Oct 1968	
Soyuz 20	17 Nov 1975	16 Feb 1976	
Soyuz 34	6 Jun 1979	19 Aug 1979	
Soyuz T1	16 Dec 1979	25 Mar 1980	
Soyuz TM1	21 May 1986	30 May 1986	

SPUTNIK (Precursors to Manned Flights)

MISSION	LAUNCH	LANDING, RECOVERY, OR RE-ENTRY	NOTES
Sputnik 1	4 Oct 1957	4 Jan 1958	
Sputnik 2	3 Nov 1957	14 Apr 1958	
Sputnik 3	15 May 1958	6 Apr 1960	
Sputnik 4	15 May 1960	5 Sep 1962	Human dummy
Sputnik 5	19 Aug 1960	20 Aug 1960	Two dogs
Sputnik 6	1 Dec 1960	2 Dec 1960	Two dogs
Sputnik 7	4 Feb 1961	26 Feb 1961	
Sputnik 8	12 Feb 1961	25 Feb 1961	Launched Venera 1
Sputnik 9	9 Mar 1961	9 Mar 1961	Dog and human dummy
Sputnik 10	25 Mar 1961	25 Mar 1961	Dog and human dummy

TABLE 2   Unmanned U.S. Missions that Achieved Earth Orbit

MISSION	LAUNCH	LANDING, RECOVERY, OR RE-ENTRY	NOTES
AC: Atlas-Centaur (Launch Vehicle)			
AC 2	27 Nov 1963		
AC 4	11 Dec 1964	12 Dec 1964	
AC 6	11 Aug 1965		
AC 8	7 Apr 1966	5 May 1966	
AC 9	25 Oct 1966	6 Nov 1966	
APOLLO TESTS (Precursors to manned missions)			
Apollo 2	5 Jul 1966	5 Jul 1966	
Apollo 4	9 Nov 1967	9 Nov 1967	
Apollo 5	22 Jan 1968	24 Jan 1968	
Apollo 6	4 Apr 1968	4 Apr 1968	
ATS: Application Technology Satellite			
ATS 1	6 Dec 1966		
ATS 2	5 Apr 1967	2 Sep 1969	
ATS 3	5 Nov 1967		
ATS 4	10 Aug 1968	17 Oct 1968	
ATS 5	12 Aug 1969		
ATS 6	30 May 1974		
BIG BIRD (Reconnaissance and Surveillance)			
Big Bird 1	15 Jun 1971	6 Aug 1971	Also known as SAMOS 83
Big Bird 2	20 Jan 1972	29 Feb 1972	Also known as SAMOS 86
Big Bird 3	7 Jul 1972	13 Sep 1972	Also known as SAMOS 88
Big Bird 4	10 Oct 1972	8 Jan 1973	Also known as SAMOS 90
Big Bird 5	9 Mar 1973	19 May 1973	Also known as SAMOS 92
Big Bird 6	13 Jul 1973	12 Oct 1973	Also known as SAMOS 94
Big Bird 7	10 Nov 1973	13 Mar 1974	Also known as SAMOS 96
Big Bird 8	10 Apr 1974	28 Jul 1974	Also known as SAMOS 98
Big Bird 9	29 Oct 1974	19 Mar 1975	Also known as SAMOS 101
Big Bird 10	8 Jun 1975	5 Nov 1975	Also known as SAMOS 103
Big Bird 11	4 Dec 1975	1 Apr 1976	
Big Bird 12	8 Jul 1976	13 Dec 1976	
Big Bird 13	27 Jun 1977	23 Dec 1977	
Big Bird 14	16 Mar 1978	11 Sep 1978	
Big Bird 15	16 Mar 1979	22 Sep 1979	
Big Bird 16	18 Jun 1980	6 Mar 1981	
Big Bird 17	28 Feb 1981	20 Jun 1981	
Big Bird 18	21 Jan 1982	23 May 1982	
Big Bird 19	11 May 1982	5 Dec 1982	
Big Bird 20	15 Apr 1983	21 Aug 1983	
Big Bird 21	20 Jun 1983	21 Mar 1984	
Big Bird 22	17 Apr 1984	13 Aug 1984	
Big Bird 23	25 Jun 1984	18 Oct 1984	Also known as USA 2

TABLE 2   89

MISSION	LAUNCH	LANDING, RECOVERY, OR RE-ENTRY	NOTES
BIOSAT (Biological Research)			
Biosat 1	14 Dec 1966	15 Feb 1967	
Biosat 2	7 Sep 1967	11 Sep 1967	
Biosat 3	28 Jun 1969	20 Jan 1970	
BMEWS: Ballistic Missile Early Warning Satellite			
BMEWS 1	6 Aug 1968		
BMEWS 2	13 Apr 1969		
BMEWS 3	19 Jun 1970		
BMEWS 4	1 Sep 1970		
BMEWS 6	20 Dec 1972		
BMEWS 7	18 Jun 1975		
CALSPHERE (Air Force Radar Calibration Satellites)			
Calsphere 1	12 Dec 1962	1 Jul 1963	
Calsphere 2	6 Oct 1964		
Calsphere 3	6 Oct 1964		
Calsphere 4	13 Aug 1965		
Calsphere 5	17 Feb 1971		
Calsphere 6	17 Feb 1971	20 Sep 1989	
Calsphere 7	17 Feb 1971		
COMSTAR (Communications)			
Comstar 1	13 May 1976		
Comstar 2	22 Jul 1976		
Comstar 3	29 Jun 1978		
Comstar 4	21 Feb 1981		
DISCOVERER (Reconnaissance and Surveillance)			
Discoverer 1	28 Feb 1959	5 Mar 1959	
Discoverer 2	13 Apr 1959	26 Apr 1959	
Discoverer 5	13 Aug 1959	28 Sep 1959	
Discoverer 6	19 Aug 1959	20 Oct 1959	
Discoverer 7	7 Nov 1959	26 Nov 1959	
Discoverer 8	20 Nov 1959	8 Mar 1960	
Discoverer 11	15 Apr 1960	26 Apr 1960	
Discoverer 13	10 Aug 1960	14 Nov 1960	
Discoverer 14	18 Aug 1960	16 Sep 1960	
Discoverer 15	13 Sep 1960	18 Oct 1960	
Discoverer 17	12 Nov 1960	29 Dec 1960	
Discoverer 18	7 Dec 1960	2 Apr 1961	
Discoverer 19	20 Dec 1960	23 Jan 1961	
Discoverer 20	17 Feb 1961	28 Jul 1962	
Discoverer 21	18 Feb 1961	20 Apr 1962	
Discoverer 23	8 Apr 1961	16 Apr 1962	
Discoverer 25	16 Jun 1961	12 Jul 1961	
Discoverer 26	7 Jul 1961	5 Dec 1961	
Discoverer 29	30 Aug 1961	10 Sep 1961	
Discoverer 30	12 Sep 1961	11 Dec 1961	
Discoverer 31	17 Sep 1961	26 Oct 1961	

MISSION	LAUNCH	LANDING, RECOVERY, OR RE-ENTRY	NOTES
Discoverer 32	13 Oct 1961	13 Nov 1961	
Discoverer 34	5 Nov 1961	7 Dec 1962	
Discoverer 35	15 Nov 1961	3 Dec 1961	
Discoverer 36	12 Dec 1961	8 Mar 1962	
Discoverer 38	27 Feb 1962	21 Mar 1962	
Discoverer 39	17 Apr 1962	28 May 1962	
Discoverer 40	28 Apr 1962	26 May 1962	
Discoverer 41	15 May 1962	26 Nov 1963	
Discoverer 42	29 May 1962	11 Jun 1962	
Discoverer 43	1 Jun 1962	28 Jun 1962	
Discoverer 44	22 Jun 1962	7 Jul 1962	
Discoverer 45	27 Jun 1962	14 Sep 1962	
Discoverer 46	20 Jul 1962	14 Aug 1962	
Discoverer 47	27 Jul 1962	24 Aug 1962	
Discoverer 48	1 Aug 1962	26 Aug 1962	
Discoverer 50	1 Sep 1962	26 Oct 1964	
Discoverer 49	10 Sep 1962	28 Aug 1986	
Discoverer 51	17 Sep 1962	16 Nov 1962	
Discoverer 52	29 Sep 1962	14 Oct 1962	
Discoverer 53	9 Oct 1962	16 Nov 1962	
Discoverer 54	5 Nov 1962	3 Dec 1962	
Discoverer 55	24 Nov 1962	13 Dec 1962	
Discoverer 56	4 Dec 1962	8 Dec 1962	
Discoverer 57	14 Dec 1962	8 Jan 1963	
Discoverer 58	7 Jan 1963	24 Jan 1963	
Discoverer 61	1 Apr 1963	26 Apr 1963	
Discoverer 63	18 May 1963	27 May 1963	
Discoverer 64	12 Jun 1963	11 Jul 1963	
Discoverer 65	26 Jun 1963	26 Jul 1963	
Discoverer 66	18 Jul 1963	13 Aug 1963	
Discoverer 67	30 Jul 1963	11 Aug 1963	
Discoverer 68	24 Aug 1963	12 Sep 1963	
Discoverer 69	29 Aug 1963	7 Nov 1963	
Discoverer 70	23 Sep 1963	12 Oct 1963	
Discoverer 71	29 Oct 1963	21 Jan 1964	
Discoverer 73	27 Nov 1963	15 Dec 1963	
Discoverer 74	21 Dec 1963	8 Jan 1964	
Discoverer 75	15 Feb 1964	9 Mar 1964	
Discoverer 77	27 Apr 1964	26 May 1964	
Discoverer 78	4 Jun 1964	18 Jun 1964	
Discoverer 79	13 Jun 1964	2 Jun 1965	
Discoverer 80	19 Jun 1964	16 Jul 1964	
Discoverer 81	10 Jul 1964	6 Aug 1964	
Discoverer 82	5 Aug 1964	31 Aug 1964	
Discoverer 83	21 Aug 1964	31 Mar 1965	
Discoverer 84	14 Sep 1964	6 Oct 1964	
Discoverer 85	5 Oct 1964	26 Oct 1964	
Discoverer 86	17 Oct 1964	4 Nov 1964	

TABLE 2   91

MISSION	LAUNCH	LANDING, RECOVERY, OR RE-ENTRY	NOTES
Discoverer 87	2 Nov 1964	28 Nov 1964	
Discoverer 88	18 Nov 1964	6 Dec 1964	
Discoverer 89	19 Dec 1964	14 Jan 1965	
Discoverer 90	15 Jan 1965	9 Feb 1965	
Discoverer 91	25 Feb 1965	18 Mar 1965	
Discoverer 92	25 Mar 1965	4 Apr 1965	
Discoverer 93	29 Apr 1965	26 May 1965	
Discoverer 94	18 May 1965	15 Jun 1965	
Discoverer 95	9 Jun 1965	22 Jun 1965	
Discoverer 96	19 Jul 1965	18 Aug 1965	
Discoverer 97	17 Aug 1965	11 Oct 1965	
Discoverer 98	22 Sep 1965	11 Oct 1965	
Discoverer 99	5 Oct 1965	29 Oct 1965	
Discoverer 100	28 Oct 1965	17 Nov 1965	
Discoverer 101	9 Dec 1965	26 Dec 1965	
Discoverer 102	24 Dec 1965	20 Jan 1966	
Discoverer 103	2 Feb 1966	27 Feb 1966	
Discoverer 104	9 Mar 1966	29 Mar 1966	
Discoverer 105	7 Apr 1966	26 Apr 1966	
Discoverer 107	23 May 1966	9 Jun 1966	
Discoverer 108	21 Jun 1966	14 Jul 1966	
Discoverer 109	9 Aug 1966	11 Sep 1966	
Discoverer 110	20 Sep 1966	12 Oct 1966	
Discoverer 111	8 Nov 1966	29 Nov 1966	
Discoverer 112	14 Jan 1967	2 Feb 1967	
Discoverer 113	22 Feb 1967	11 Mar 1967	
Discoverer 114	30 Mar 1967	17 Apr 1967	
Discoverer 115	9 May 1967	13 Jul 1967	
Discoverer 116	16 Jun 1967	20 Jul 1967	
Discoverer 117	7 Aug 1967	1 Sep 1967	
Discoverer 118	15 Sep 1967	4 Oct 1967	
Discoverer 119	2 Nov 1967	2 Dec 1967	
Discoverer 120	9 Dec 1967	25 Dec 1967	
Discoverer 121	24 Jan 1968	27 Feb 1968	
Discoverer 122	14 Mar 1968	10 Apr 1968	
Discoverer 123	1 May 1968	15 May 1968	
Discoverer 124	20 Jun 1968	16 Jul 1968	
Discoverer 125	7 Aug 1968	27 Aug 1968	
Discoverer 126	18 Sep 1968	8 Oct 1968	
Discoverer 127	3 Nov 1968	23 Nov 1968	
Discoverer 128	12 Dec 1968	28 Dec 1968	
Discoverer 129	5 Feb 1969	24 Feb 1969	
Discoverer 130	19 Mar 1969	24 Mar 1969	
Discoverer 131	1 May 1969	23 May 1969	
Discoverer 132	23 Jul 1969	23 Aug 1969	
Discoverer 133	22 Sep 1969	12 Oct 1969	
Discoverer 134	4 Dec 1969	10 Jan 1970	
Discoverer 135	4 Mar 1970	26 Mar 1970	

MISSION	LAUNCH	LANDING, RECOVERY, OR RE-ENTRY	NOTES
Discoverer 136	20 May 1970	17 Jun 1970	
Discoverer 137	23 Jul 1970	19 Aug 1970	
Discoverer 138	18 Nov 1970	11 Dec 1970	
Discoverer 140	24 Mar 1971	12 Apr 1971	
Discoverer 141	10 Sep 1971	5 Oct 1971	
Discoverer 142	19 Apr 1972	12 May 1972	
Discoverer 143	25 May 1972	4 Jun 1972	

DMSP: Defense Meteorological Satellite Program

DMSP 5D 1	24 May 1975		
DMSP 5D 2	11 Sep 1976		
DMSP 5D 3	5 Jun 1977		
DMSP 5D 4	1 May 1978		
DMSP 5D 5	6 Jun 1979		
DMSP 5D 6	21 Dec 1982		
DMSP 5D 7	18 Nov 1983		
DMSP 5D 8	20 Jun 1987		Also known as USA 26

DSCS: Defense Satellite Communications System

DSCS II 1	3 Nov 1971		
DSCS II 2	3 Nov 1971		
DSCS II 3	13 Dec 1973		
DSCS II 4	13 Dec 1973		
DSCS II 5	20 May 1975	26 May 1975	
DSCS II 6	20 May 1975	26 May 1975	
DSCS II 7	12 May 1977		
DSCS II 8	12 May 1977		
DSCS II 11	14 Dec 1978		
DSCS II 12	14 Dec 1978		
DSCS II 13	21 Nov 1979		
DSCS II 14	21 Nov 1979		
DSCS II 15	30 Oct 1982		
DSCS II 16	30 Oct 1982		Also known as DSCS III 1
DSCS III 1	30 Oct 1982		Also known as DSCS II 16
DSCS III 2	31 Jan 1984		
DSCS III 3	3 Oct 1985		Also known as USA 11
DSCS III 4	3 Oct 1985		Also known as USA 12

ECHO (Communications)

Echo 1	12 Aug 1960	24 May 1968	
Echo 2	25 Jan 1964	7 Jun 1969	

ERS: Environmental Research Satellite

ERS 2	17 Sep 1962	16 Nov 1962	Tetrahedral Research Satellite, TRS 1
ERS 5	9 May 1963		Tetrahedral Research Satellite, TRS 2
ERS 6	9 May 1963		Tetrahedral Research Satellite, TRS 3
ERS 9	18 Jul 1963		Tetrahedral Research Satellite, TRS 4

TABLE 2   93

MISSION	LAUNCH	LANDING, RECOVERY, OR RE-ENTRY	NOTES
ERS 10	18 Jul 1963		
ERS 12	16 Oct 1963	1 Jul 1965	Tetrahedral Research Satellite, TRS 5
ERS 13	17 Jul 1964	1 Jul 1966	Tetrahedral Research Satellite, TRS 6
ERS 17	20 Jul 1965	1 Jul 1968	Octahedral Research Satellite, ORS 3
ERS 16	9 Jun 1966	12 Mar 1967	Octahedral Research Satellite, ORS 2
ERS 15	19 Aug 1966		Octahedral Research Satellite, ORS 1
ERS 18	28 Apr 1967		
ERS 20	28 Apr 1967		Also known as OV5-3
ERS 27	28 Apr 1967		Also known as OV5-1
ERS 30	13 Dec 1967	28 Apr 1968	Also known as TETR 1
ERS 21	26 Sep 1968		Also known as OV5-4
ERS 28	26 Sep 1968	15 Feb 1971	Also known as OV5-2
ERS 29	23 May 1969		Also known as OV5-5

ESSA: Environmental Sciences Services Administration

MISSION	LAUNCH	LANDING, RECOVERY, OR RE-ENTRY	NOTES
ESSA 1	3 Feb 1966		
ESSA 2	28 Feb 1966		
ESSA 3	2 Oct 1966		
ESSA 4	26 Jan 1967		
ESSA 5	20 Apr 1967		
ESSA 6	10 Nov 1967		
ESSA 7	16 Aug 1968		
ESSA 8	15 Dec 1968		
ESSA 9	26 Feb 1969		

EXPLORER (Science, Technology)

MISSION	LAUNCH	LANDING, RECOVERY, OR RE-ENTRY	NOTES
Explorer 1	31 Jan 1958	31 Mar 1970	
Explorer 3	26 Mar 1958	28 Jun 1958	
Explorer 4	26 Jul 1958	23 Oct 1959	
Explorer 6	7 Aug 1959	Jul 1961	
Explorer 7	13 Oct 1959		
Explorer 8	3 Nov 1960		
Explorer 9	16 Feb 1961	9 Apr 1964	
Explorer 10	25 Mar 1961	Jun 1968	
Explorer 11	27 Apr 1961		
Explorer 12	15 Aug 1961	Sep 1963	
Explorer 13	25 Aug 1961	28 Aug 1961	
Explorer 14	2 Oct 1962	1 Jul 1966	
Explorer 15	27 Oct 1962	19 Dec 1983	
Explorer 16	16 Dec 1962		
Explorer 17	2 Apr 1963	24 Nov 1966	Atmospheric Explorer, AE 1
Explorer 18	26 Nov 1963	Dec 1965	Interplanetary Monitoring Platform, IMP-A
Explorer 19	19 Dec 1963	10 May 1981	
Explorer 20	25 Aug 1964		

MISSION	LAUNCH	LANDING, RECOVERY, OR RE-ENTRY	NOTES
Explorer 21	3 Oct 1964	Jan 1966	Interplanetary Monitoring Platform, IMP-B
Explorer 22	9 Oct 1964		
Explorer 23	6 Nov 1964	29 Jun 1983	
Explorer 24	21 Nov 1964	18 Oct 1968	
Explorer 25	21 Nov 1964		Also known as Injun 4
Explorer 26	21 Dec 1964		
Explorer 27	29 Apr 1965		
Explorer 28	29 May 1965	4 Jul 1968	Interplanetary Monitoring Platform, IMP-C
Explorer 29	6 Nov 1965		Also known as GEOS 1
Explorer 30	19 Nov 1965		Also known as Solrad 8
Explorer 31	28 Nov 1965		
Explorer 32	25 May 1966	22 Feb 1985	Atmospheric Explorer, AE 2
Explorer 33	1 Jul 1966		Interplanetary Monitoring Platform, IMP-D
Explorer 34	24 May 1967	3 May 1969	Interplanetary Monitoring Platform, IMP-F
Explorer 35	19 Jul 1967		Interplanetary Monitoring Platform, IMP-E
Explorer 36	11 Jan 1968		Also known as GEOS 2
Explorer 37	5 Mar 1968		Also known as Solrad 9
Explorer 38	4 Jul 1968		Radio Astronomy Explorer, RAE 1
Explorer 39	8 Aug 1968	22 Jun 1981	
Explorer 40	8 Aug 1968		Also known as Injun 5
Explorer 41	21 Jun 1969	23 Dec 1972	Interplanetary Monitoring Platform, IMP-G
Explorer 42	12 Dec 1970	5 Apr 1979	Small Astronomical Satellite, SAS 1
Explorer 43	13 Mar 1971	2 Oct 1974	Interplanetary Monitoring Platform, IMP-I
Explorer 44	8 Jul 1971	15 Dec 1979	Also known as Solrad 10
Explorer 45	15 Nov 1971		
Explorer 46	13 Aug 1972	2 Nov 1979	
Explorer 47	23 Sep 1972		Interplanetary Monitoring Platform, IMP-H
Explorer 48	15 Nov 1972	1 May 1979	Small Astronomical Satellite, SAS 2
Explorer 49	10 Jun 1973		Radio Astronomy Explorer, RAE 2
Explorer 50	25 Oct 1973		Interplanetary Monitoring Platform, IMP-J
Explorer 51	13 Dec 1973	12 Dec 1978	Atmospheric Explorer, AE 3
Explorer 52	3 Jun 1974	28 Apr 1978	
Explorer 53	7 May 1975	9 Apr 1979	Small Astronomical Satellite, SAS 3
Explorer 54	6 Oct 1975	12 Mar 1976	Atmospheric Explorer, AE 4
Explorer 55	20 Nov 1975	10 Jun 1981	Atmospheric Explorer, AE 5

FERRET (Reconnaissance and Surveillance)

Ferret 1	21 Feb 1962	4 Mar 1962	

TABLE 2   95

MISSION	LAUNCH	LANDING, RECOVERY, OR RE-ENTRY	NOTES
Ferret 2	18 Jun 1962	29 Oct 1963	
Ferret 3	12 Dec 1962	9 Feb 1967	
Ferret 4	16 Jan 1963	9 Jan 1969	
Ferret 5	15 Jun 1963	27 Jul 1963	
Ferret 6	29 Jun 1963	26 Oct 1969	
Ferret 7	11 Jan 1964		
Ferret 8	27 Feb 1964	19 Feb 1969	
Ferret 9	2 Jul 1964	8 Jul 1969	
Ferret 10	3 Nov 1964	5 Nov 1969	
Ferret 11	21 Dec 1964	11 Jan 1965	
Ferret 12	9 Mar 1965		Also known as Solrad 6B
Ferret 13	16 Jul 1965	18 Dec 1968	
Ferret 14	9 Feb 1966	26 Sep 1969	
Ferret 15	28 Dec 1966	5 Apr 1969	
Ferret 16	31 May 1967		
Ferret 17	24 Jul 1967	5 Jun 1969	
Ferret 18	17 Jan 1968	7 Jul 1970	
Ferret 19	5 Oct 1968	26 Mar 1971	
Ferret 20	31 Jul 1969	4 Jan 1973	
Ferret 21	30 Sep 1969		
Ferret 22	26 Aug 1970	26 Mar 1975	
Ferret 23	16 Jul 1971	31 Aug 1978	
Ferret 24	14 Dec 1971		

FLTSATCOM (Naval Communications)

FLTSATCOM 1	9 Feb 1978		
FLTSATCOM 2	4 May 1979		
FLTSATCOM 3	18 Jan 1980		
FLTSATCOM 4	31 Oct 1980		
FLTSATCOM 5	6 Aug 1981		
FLTSATCOM 7	5 Dec 1986		Also known as USA 20

G STAR (Communications)

G Star 1	8 May 1985		
G Star 2	28 Mar 1986		

GALAXY (Communications)

Galaxy 1	28 Jun 1983		
Galaxy 2	22 Sep 1983		
Galaxy 3	21 Sep 1984		

GEMINI TESTS (Docking Tests)

Gemini 1	8 Apr 1964	12 Apr 1964	
G 8 Target	16 Mar 1966	15 Sep 1967	
G 9 Target B	1 Jun 1966	11 Jul 1966	
G 10 Target	18 Jul 1966	29 Dec 1966	
G 11 Target	12 Sep 1966	30 Dec 1966	
G 12 Target	11 Nov 1966	23 Dec 1966	

GGSE: Gravity Gradient Stabilization Experiment

GGSE 1	11 Jan 1964		

MISSION	LAUNCH	LANDING, RECOVERY, OR RE-ENTRY	NOTES
GGSE 2	9 Mar 1965		
GGSE 3	9 Mar 1965		
GGSE 4	31 May 1967		
GGSE 5	31 May 1967		

GOES: Geostationary Operational Environmental Satellite

GOES 1	16 Oct 1975		Also known as SMS-C
GOES 2	16 Jun 1977		
GOES 3	16 Jun 1978		
GOES 4	9 Sep 1980		
GOES 5	22 May 1981		
GOES 6	28 Apr 1983		
GOES 7	26 Feb 1987		

HEAO: High Energy Astronomy Observatory

HEAO 1	12 Aug 1977	15 Mar 1979	
HEAO 2	13 Nov 1978	25 Mar 1982	
HEAO 3	20 Sep 1979	7 Dec 1981	

HITCHHIKER (Military Applications)

Hitchhiker 2	26 Jun 1963	
Hitchhiker 3	29 Oct 1963	23 May 1965
Hitchhiker 4	21 Dec 1963	7 Nov 1964
Hitchhiker 5	6 Jul 1964	3 Jan 1965
Hitchhiker 6	14 Aug 1964	8 Mar 1979
Hitchhiker 7	23 Oct 1964	23 Feb 1965
Hitchhiker 8	28 Apr 1965	31 Oct 1969
Hitchhiker 9	25 Jun 1965	22 Aug 1968
Hitchhiker 10	3 Aug 1965	17 Jun 1968
Hitchhiker 11	14 May 1966	27 Oct 1970
Hitchhiker 12	16 Aug 1966	5 Mar 1970
Hitchhiker 13	16 Sep 1966	9 May 1968
Hitchhiker 14	9 May 1967	
Hitchhiker 15	16 Jun 1967	22 Oct 1968
Hitchhiker 16	2 Nov 1967	28 Mar 1969
Hitchhiker 17	24 Jan 1968	4 Mar 1970
Hitchhiker 18	14 Mar 1968	3 Jan 1970
Hitchhiker 19	20 Jun 1968	11 Jan 1970
Hitchhiker 20	18 Sep 1968	28 Sep 1969
Hitchhiker 21	12 Dec 1968	
Hitchhiker 22	5 Feb 1969	
Hitchhiker 23	19 Mar 1969	6 Dec 1971
Hitchhiker 24	1 May 1969	16 Feb 1970
Hitchhiker 25	22 Sep 1969	16 May 1971
Hitchhiker 26	4 Mar 1970	10 Nov 1971
Hitchhiker 27	20 May 1970	8 Mar 1974
Hitchhiker 28	18 Nov 1970	14 Sep 1977
Hitchhiker 29	10 Sep 1971	3 Feb 1976
Hitchhiker 30	20 Jan 1972	23 Jan 1972
Hitchhiker 31	7 Jul 1972	6 May 1978

TABLE 2 97

MISSION	LAUNCH	LANDING, RECOVERY, OR RE-ENTRY	NOTES
Hitchhiker 32	10 Oct 1972		
Hitchhiker 33	10 Nov 1973	26 Dec 1978	
Hitchhiker 34	10 Nov 1973	13 Nov 1973	
Hitchhiker 35	10 Apr 1974		
Hitchhiker 36	10 Apr 1974	22 Feb 1980	
Hitchhiker 37	29 Oct 1974	23 Jan 1980	
Hitchhiker 38	4 Dec 1975	1 May 1978	
Hitchhiker 39	8 Jul 1976	24 Apr 1986	
Hitchhiker 40	16 Mar 1978		
Hitchhiker 41	16 Mar 1979		
Hitchhiker 42	11 May 1982		

IDCSP: Initial Defense Communications Satellite Program

IDCSP 1	16 Jun 1966
IDCSP 2	16 Jun 1966
IDCSP 3	16 Jun 1966
IDCSP 4	16 Jun 1966
IDCSP 5	16 Jun 1966
IDCSP 6	16 Jun 1966
IDCSP 7	16 Jun 1966
IDCSP 8	18 Jan 1967
IDCSP 9	18 Jan 1967
IDCSP 10	18 Jan 1967
IDCSP 11	18 Jan 1967
IDCSP 12	18 Jan 1967
IDCSP 13	18 Jan 1967
IDCSP 14	18 Jan 1967
IDCSP 15	18 Jan 1967
IDCSP 16	1 Jul 1967
IDCSP 17	1 Jul 1967
IDCSP 18	1 Jul 1967
IDCSP 19	13 Jun 1968
IDCSP 20	13 Jun 1968
IDCSP 21	13 Jun 1968
IDCSP 22	13 Jun 1968
IDCSP 23	13 Jun 1968
IDCSP 24	13 Jun 1968
IDCSP 25	13 Jun 1968
IDCSP 26	13 Jun 1968

IMEWS: Integrated Missile Early Warning Satellite

IMEWS 1	6 Nov 1970
IMEWS 2	5 May 1971
IMEWS 3	1 Mar 1972
IMEWS 4	12 Jun 1973
IMEWS 5	14 Dec 1975
IMEWS 6	26 Jun 1976
IMEWS 7	6 Feb 1977
IMEWS 8	10 Jun 1978

MISSION	LAUNCH	LANDING, RECOVERY, OR RE-ENTRY	NOTES
IMEWS 9	10 Jun 1979		
IMEWS 10	1 Oct 1979		
IMEWS 11	16 Mar 1981		
IMEWS 12	31 Oct 1981		
IMEWS 13	6 Mar 1982		
IMEWS 14	14 Apr 1984		
IMEWS 15	22 Dec 1984		Also known as USA 7
IMEWS 16	29 Nov 1987		Also known as USA 28

INJUN (Science, Technology)

MISSION	LAUNCH	LANDING, RECOVERY, OR RE-ENTRY	NOTES
Injun 1	29 Jun 1961		
Injun 3	12 Dec 1962	25 Aug 1968	
Injun 4	21 Nov 1964		Also known as Explorer 25
Injun 5	8 Aug 1968		Also known as Explorer 40

KH (Reconnaissance and Surveillance)

MISSION	LAUNCH	LANDING, RECOVERY, OR RE-ENTRY	NOTES
KH 11-1	19 Dec 1976	28 Jan 1979	
KH 11-2	14 Jun 1978	23 Aug 1981	
KH 11-3	7 Feb 1980	30 Oct 1982	
KH 11-4	3 Sep 1981	23 Nov 1984	
KH 11-5	17 Nov 1982	13 Aug 1985	
KH 11-6	4 Dec 1984		Also known as USA 6
KH 11-7	26 Oct 1987		Also known as USA 27

LANDSAT (Earth Observation)

MISSION	LAUNCH	LANDING, RECOVERY, OR RE-ENTRY	NOTES
Landsat 1	23 Jul 1972		
Landsat 2	22 Jan 1975		
Landsat 3	5 Mar 1978		
Landsat 4	16 Jul 1982		
Landsat 5	1 Mar 1984		

LCS: Lincoln Calibration Sphere

MISSION	LAUNCH	LANDING, RECOVERY, OR RE-ENTRY	NOTES
LCS 1	6 May 1965		
LCS 2	15 Oct 1965	27 Jul 1972	
LCS 4	7 Aug 1971	1 Sep 1981	Also known as Rigid Sphere 1

LES: Lincoln Experimental Satellite

MISSION	LAUNCH	LANDING, RECOVERY, OR RE-ENTRY	NOTES
LES 1	11 Feb 1965		
LES 2	6 May 1965		
LES 3	21 Dec 1965	6 Apr 1968	
LES 4	21 Dec 1965	1 Aug 1977	
LES 5	1 Jul 1967		
LES 6	26 Sep 1968		
LES 8	15 Mar 1976		
LES 9	15 Mar 1976		

LOFTI: Low Frequency Trans Ionospheric Satellite

MISSION	LAUNCH	LANDING, RECOVERY, OR RE-ENTRY	NOTES
LOFTI 1	21 Feb 1961	30 Mar 1961	
LOFTI 2	15 Jun 1963	18 Jul 1963	

TABLE 2   99

MISSION	LAUNCH	LANDING, RECOVERY, OR RE-ENTRY	NOTES
MERCURY TESTS (Precursors to Manned Flights)			
Big Joe	9 Sep 1959		
LJ 6	4 Oct 1959		
LJ 2	4 Dec 1959		Monkey passenger
LJ 1B	21 Jan 1960		Monkey passenger
MR 1A	19 Dec 1960		
MR 2	21 Feb 1961		Monkey passenger
MR BD	24 Mar 1961		
MA 4	13 Sep 1961	13 Sep 1961	
MA 5	29 Nov 1961	29 Nov 1961	Monkey passenger
MIDAS: Missile Defense Alarm System			
MIDAS 2	24 May 1960	7 Feb 1974	
MIDAS 3	12 Jul 1961		
MIDAS 4	21 Oct 1961		
MIDAS 5	9 Apr 1962		
MIDAS 7	9 May 1963		
MIDAS 9	18 Jul 1963		
I. MIDAS 1	9 Jun 1966	3 Dec 1966	Improved MIDAS
I. MIDAS 2	19 Aug 1966		
I. MIDAS 3	5 Oct 1966		
NAVSTAR (Navigation, U.S. Navy)			
Navstar 1	22 Feb 1978		
Navstar 2	13 May 1978		
Navstar 3	7 Oct 1978		
Navstar 4	11 Dec 1978		
Navstar 5	9 Feb 1980		
Navstar 6	26 Apr 1980		
Navstar 8	14 Jul 1983		
Navstar 9	13 Jun 1984		Also known as USA 1
Navstar 10	8 Sep 1984		Also known as USA 5
Navstar 11	9 Oct 1985		Also known as USA 10
NIMBUS (Meteorological)			
Nimbus 1	28 Aug 1964	16 May 1974	
Nimbus 2	15 May 1966		
Nimbus 3	14 Apr 1969		
Nimbus 4	8 Apr 1970		
Nimbus 5	11 Dec 1972		
Nimbus 6	12 Jun 1975		
Nimbus 7	24 Oct 1978		
NNSS: Navy Navigational Satellite System			
NNSS 30010	6 Oct 1964		
NNSS 30020	12 Dec 1964		
NNSS 30030	11 Mar 1965	14 Jun 1965	
NNSS 30040	24 Jun 1965		
NNSS 30050	13 Aug 1965		

MISSION	LAUNCH	LANDING, RECOVERY, OR RE-ENTRY	NOTES
NNSS 30060	21 Dec 1965		
NNSS 30070	28 Jan 1966		
NNSS 30080	25 Mar 1966		
NNSS 30090	19 May 1966		
NNSS 30100	17 Aug 1966		
NNSS 30120	13 Apr 1967		
NNSS 30130	18 May 1967		
NNSS 30140	25 Sep 1967		
NNSS 30180	1 Mar 1968		
NNSS 30190	27 Aug 1970		
NNSS 30200	30 Oct 1973		
NNSS 30240	3 Aug 1985		
NNSS 30300	3 Aug 1985		
NNSS 30270	16 Sep 1987		
NNSS 30290	16 Sep 1987		

NOAA: National Oceanics and Atmospheric Administration

MISSION	LAUNCH	LANDING, RECOVERY, OR RE-ENTRY	NOTES
NOAA 1	11 Dec 1970		
NOAA 2	15 Oct 1972		
NOAA 3	6 Nov 1973		
NOAA 4	15 Nov 1974		
NOAA 5	29 Jul 1976		
NOAA 6	27 Jun 1979		
NOAA B	29 May 1980	3 May 1981	
NOAA 7	23 Jun 1981		
NOAA 8	28 Mar 1983		
NOAA 9	12 Dec 1984		
NOAA 10	17 Sep 1986		

NOSS: Naval Ocean Surveillance Satellite

MISSION	LAUNCH	LANDING, RECOVERY, OR RE-ENTRY	NOTES
NOSS 1	30 Apr 1976		
NOSS 2	8 Dec 1977		
NOSS 3	3 Mar 1980		
NOSS 4	9 Feb 1983		
NOSS 5	9 Jun 1983		
NOSS 6	5 Feb 1984		
NOSS 7	9 Feb 1986		Also known as USA 15
NOSS 8	15 May 1987		Also known as USA 22

NOVA (Navigation, U.S. Navy)

MISSION	LAUNCH	LANDING, RECOVERY, OR RE-ENTRY	NOTES
Nova 1	15 May 1981		
Nova 2	11 Oct 1984		

OAO: Orbiting Astronomical Observatory

MISSION	LAUNCH	LANDING, RECOVERY, OR RE-ENTRY	NOTES
OAO 1	8 Apr 1966		
OAO 2	7 Dec 1968		
OAO 3	21 Aug 1972		

OGO: Orbiting Geophysical Observatory

MISSION	LAUNCH	LANDING, RECOVERY, OR RE-ENTRY	NOTES
OGO 1	4 Sep 1964		
OGO 2	14 Oct 1965	17 Sep 1983	

TABLE 2   101

MISSION	LAUNCH	LANDING, RECOVERY, OR RE-ENTRY	NOTES
OGO 3	6 Jun 1966		
OGO 4	28 Jul 1967	16 Aug 1972	
OGO 5	4 Mar 1968		
OGO 6	5 Jun 1969	12 Oct 1979	
OSO: Orbiting Solar Observatory			
OSO 1	7 Mar 1962	8 Oct 1981	
OSO 2	3 Feb 1965	9 Aug 1989	
OSO 3	8 Mar 1967	4 Apr 1982	
OSO 4	18 Oct 1967	15 Jun 1982	
OSO 5	22 Jan 1969	2 Apr 1984	
OSO 6	9 Aug 1969	7 Mar 1981	
OSO 7	29 Sep 1971	9 Jul 1974	
OSO 8	21 Jun 1975	9 Jul 1986	
OTHER SATELLITES			
SCORE	18 Dec 1958	21 Jan 1959	Signal Corps Orbiting Relay Experiment
Courier 1B	4 Oct 1960		
TRAAC	15 Nov 1961		Transit Research and Attitude Control
TAVE	29 Sep 1962		Thor-Agena Vibration Experiment
Starad 1	26 Oct 1962	5 Oct 1967	
ANNA 1B	31 Oct 1962		Army, Navy, NASA, Air Force
GRS	28 Jun 1963	14 Dec 1983	Geophysical Research Satellite
DASH 2	18 Jul 1963	12 Apr 1971	Density and Scale Height
Target $0.1m^2$	29 Aug 1963	28 Sep 1963	
Snapshot	3 Apr 1965		
Porcupine 2	13 Aug 1965		
Spasurrod 1	13 Aug 1965		
Tempsat 1	13 Aug 1965		
REP	21 Aug 1965	27 Aug 1965	Rendezvous Evaluation Package
Bluebell	15 Feb 1966	16 Feb 1966	
Bluebell	15 Feb 1966	22 Feb 1966	
A3	18 Mar 1966	23 Mar 1966	
GGTS	16 Jun 1966		Gravity Gradient Test Satellite
Pageos	23 Jun 1966		
SGLS	12 Oct 1966	21 Oct 1966	
LOGACS	22 May 1967	27 May 1967	Low Gravity Accelerometer Calibration System
Aurora 1	29 Jun 1967		
DATS 1	1 Jul 1967		De-spun Antenna Test Satellite
DODGE	1 Jul 1967		Department of Defense Gravity Experiment
TACSAT 1	9 Feb 1969		Tactical Communications Satellite
Orbis Cal 2	17 Mar 1969	24 Mar 1969	

MISSION	LAUNCH	LANDING, RECOVERY, OR RE-ENTRY	NOTES
PAC 1	9 Aug 1969	28 Apr 1977	Package Attitude Control
SERT 2	4 Feb 1970		Space Electric Rocket Test
Topo 1	8 Apr 1970		
OFO 1	9 Nov 1970	9 May 1971	Orbiting Frog Otholith
RM 9	9 Nov 1970	7 Feb 1971	Radiation Meteoroid
CEP 1	11 Dec 1970		Cylindrical Electron Probe
SESP 1	8 Jun 1971	31 Jan 1982	Space Experiment Support Program
Cannonball 2	7 Aug 1971	31 Jan 1972	
Grid Sphere 1	7 Aug 1971	2 Nov 1979	
Grid Sphere 2	7 Aug 1971	14 Apr 1979	
Musketball	7 Aug 1971	19 Sep 1971	
Mylar balloon	7 Aug 1971	11 Jun 1972	
Rigid Sphere 1	7 Aug 1971	1 Sep 1981	Also known as LCS 4
Rigid Sphere 2	7 Aug 1971		
STP	17 Oct 1971		Space Test Program
ITOS-B	21 Oct 1971	21 Jul 1972	
Skylab 1	14 May 1973	11 Jul 1979	
SESP 73-5	29 Oct 1974	26 May 1975	
GEOS 3	9 Apr 1975		Geodetic Earth Orbit Satellite
LAGEOS	4 May 1976		Laser Geodetic Satellite
P76 5	22 May 1976		
SESP 74-2	8 Jul 1976	24 Apr 1986	
NTS 2	23 Jun 1977		NTS 1 same as Timation 3
Transat	28 Oct 1977		
IUE	26 Jan 1978		International Ultraviolet Explorer
PIX 1	5 Mar 1978		Plasma Interaction Experiment
HCMM	26 Apr 1978	22 Dec 1981	Heat Capacity Mapping Mission
Seasat	27 Jun 1978		
CAMEO	24 Oct 1978		Chemically Active Materials Ejected in Orbit
SCATHA	30 Jan 1979		Spacecraft Charging at High Altitudes
SAGE	18 Feb 1979		Stratospheric Aerosol and Gas Experiment
Solwind P78-1	24 Feb 1979	13 Sep 1985	
Magsat	30 Oct 1979	11 Jun 1980	
SMM	14 Feb 1980		Solar Maximum Mission
DE 1	3 Aug 1981		Dynamics Explorer
DE 2	3 Aug 1981	19 Feb 1983	Dynamics Explorer
SME	6 Oct 1981		Solar Mesosphere Explorer
PIX 2	26 Jan 1983		Plasma Interaction Experiment
SS-A	9 Feb 1983		
SS-B	9 Feb 1983		
SS-C	9 Feb 1983		

TABLE 2   103

MISSION	LAUNCH	LANDING, RECOVERY, OR RE-ENTRY	NOTES
SS-D	9 Feb 1983		
TDRS 1	5 Apr 1983		Tracking and Data Relay Satellite
GB 1	9 Jun 1983		
GB 2	9 Jun 1983		
GB 3	9 Jun 1983		
HILAT 1	27 Jun 1983		High Latitude
JD 1	5 Feb 1984		
JD 2	5 Feb 1984		
JD 3	5 Feb 1984		
IRT	5 Feb 1984	11 Feb 1984	Inflatable Radar Test
LDEF	7 Apr 1984		Long Duration Exposure Facility
AMPTE—CCE	16 Aug 1984		Active Magnetospheric Particle Tracer Experiment—Charge Composition Explorer
ERBS	5 Oct 1984		Earth Radiation Budget Satellite
Geosat	13 Mar 1985		
NUSAT 1	29 Apr 1985	15 Dec 1986	Northern Utah Satellite
PDP 2	1 Aug 1985	1 Aug 1985	Plasma Diagnostic Package
ASC 1	27 Aug 1985		
GLOMR	1 Nov 1985	26 Dec 1986	Global Low Orbiting Message Relay
OEX Target	30 Nov 1985	2 Mar 1987	Orbiter Experiment
ITV 1	13 Dec 1985		Instrumented Target Vehicle; also known as USA 13
ITV 2	13 Dec 1985	9 Aug 1987	Instrumented Target Vehicle; also known as USA 14
Polar Bear	14 Nov 1986		
OV: Orbiting Vehicle			
OV1-2	5 Oct 1965		
OV2-1	15 Oct 1965	27 Jul 1972	
OV2-3	21 Dec 1965	17 Aug 1975	
OV1-4	30 Mar 1966		
OV1-5	30 Mar 1966		
OV3-1	22 Apr 1966		
OV3-4	10 Jun 1966		
OV1-8	13 Jul 1966	4 Jan 1978	
OV3-3	4 Aug 1966		
OV3-2	28 Oct 1966	29 Sep 1971	
OV1-6	3 Nov 1966	31 Dec 1966	
OV4-1R	3 Nov 1966	5 Jan 1967	
OV4-1T	3 Nov 1966	11 Jan 1967	
OV4-3	3 Nov 1966	9 Jan 1967	
OV1-9	11 Dec 1966		
OV1-10	11 Dec 1966		
OV5-1	28 Apr 1967		

MISSION	LAUNCH	LANDING, RECOVERY, OR RE-ENTRY	NOTES
OV5-3	28 Apr 1967		
OV1-12	27 Jul 1967	22 Jul 1980	
OV1-86	27 Jul 1967	22 Feb 1972	
OV3-6	4 Dec 1967	9 Mar 1969	
OV1-13	6 Apr 1968		
OV1-14	6 Apr 1968		
OV1-15	11 Jul 1968	6 Nov 1968	
OV1-16	11 Jul 1968	19 Aug 1968	
OV2-5	26 Sep 1968		
OV5-2	26 Sep 1968	15 Feb 1971	
OV5-4	26 Sep 1968		
OV1-17	17 Mar 1969	5 Mar 1970	
OV1-18	17 Mar 1969	27 Aug 1972	
OV1-19	17 Mar 1969		
OV5-5	23 May 1969		
OV5-6	23 May 1969		
OV5-9	23 May 1969		
OV1-20	7 Aug 1971	28 Aug 1971	
OV1-21	7 Aug 1971		
P35 (Military, Meteorological)			
P35-2	23 Aug 1962		
P35-3	19 Feb 1963	26 Dec 1979	
P35-6	19 Jan 1964		
P35-7	19 Jan 1964		
P35-8	17 Jun 1964		
P35-9	17 Jun 1964		
P35-10	18 Jan 1965	13 Jul 1979	
P35-11	18 Mar 1965		
P35-12	20 May 1965		
P35-13	9 Sep 1965		
P35-15	30 Mar 1966		
P35-16	15 Sep 1966		
P35-17	8 Feb 1967		
P35-18	22 Aug 1967		
P35-19	11 Oct 1967		
P35-20	22 May 1968		
P35-21	22 Oct 1968		
P35-22	22 Jul 1969		
P35-23	11 Feb 1970		
P35-24	3 Sep 1970		
P35-25	17 Feb 1971		
PEGASUS (Science, Technology)			
Pegasus 1	16 Feb 1965	17 Sep 1978	
Pegasus 2	25 May 1965	3 Nov 1979	
Pegasus 3	30 Jul 1965	4 Aug 1969	
PICKABACK (Military Applications)			
Pickaback	25 Oct 1963	28 Oct 1963	

TABLE 2   105

MISSION	LAUNCH	LANDING, RECOVERY, OR RE-ENTRY	NOTES
Pickaback	23 Oct 1964	29 Oct 1964	
Pickaback	8 Nov 1965	11 Nov 1965	
Pickaback	19 Jan 1966	23 Jan 1966	
Pickaback	3 Jun 1966	9 Jun 1966	
Pickaback	2 Nov 1966	16 Nov 1966	

RADOSE (U.S. Air Force, Science, Technology)

Radose	15 Jun 1963	30 Jul 1963	
Radose 5E1	28 Sep 1963		
Radose 5E1A	5 Dec 1963		
Radose 5E3	5 Dec 1963		
Radose 5E5	12 Dec 1964		

RANGER (Lunar Probes)

Ranger 1	23 Aug 1961	30 Aug 1961	Lunar probe but remained in Earth orbit
Ranger 2	18 Nov 1961	20 Nov 1961	Lunar probe but remained in Earth orbit

RCA AMERICOM (Communications)

RCA Americom 2	28 Nov 1985	
RCA Americom 1	12 Jan 1986	

RCA BLOCK 5B-C (Earth Observation)

Unnamed	14 Oct 1971	
Unnamed	24 Mar 1972	
Unnamed	9 Nov 1972	
Unnamed	17 Aug 1973	
Unnamed	16 Mar 1974	
Unnamed	9 Aug 1974	
Unnamed	19 Feb 1976	19 Feb 1976

RCA SATCOM (Communications)

RCA Satcom 1	13 Dec 1975
RCA Satcom 2	26 Mar 1976
RCA Satcom 3	7 Dec 1979
RCA Satcom 3R	20 Nov 1981
RCA Satcom 4	16 Jan 1982
RCA Satcom 5	20 Oct 1982
RCA Satcom 6	11 Apr 1983
RCA Satcom 7	8 Sep 1983

RELAY (Communications)

Relay 1	13 Dec 1962
Relay 2	21 Jan 1964

RH: Rhyolite Satellites (Reconnaissance and Surveillance)

RH 1	6 Mar 1973
RH 2	23 May 1977
RH 3	11 Dec 1977
RH 4	7 Apr 1978

MISSION	LAUNCH	LANDING, RECOVERY, OR RE-ENTRY	NOTES
SAMOS: Satellite and Missile Observation System			
SAMOS 2	31 Jan 1961	21 Oct 1973	
SAMOS 5	22 Dec 1961	14 Aug 1962	
SAMOS 6	7 Mar 1962	7 Jun 1963	
SAMOS 7	26 Apr 1962	28 Apr 1962	
SAMOS 8	17 Jun 1962	18 Jun 1962	
SAMOS 9	18 Jul 1962	25 Jul 1962	
SAMOS 10	5 Aug 1962	6 Aug 1962	
SAMOS 11	11 Nov 1962	12 Nov 1962	
I. SAMOS 1	12 Jul 1963	18 Jul 1963	Improved SAMOS
I. SAMOS 2	6 Sep 1963	13 Sep 1963	
I. SAMOS 3	25 Oct 1963	29 Oct 1963	
I. SAMOS 4	18 Dec 1963	20 Dec 1963	
I. SAMOS 5	25 Feb 1964	1 Mar 1964	
I. SAMOS 6	11 Mar 1964	16 Mar 1964	
I. SAMOS 7	23 Apr 1964	28 Apr 1964	
I. SAMOS 8	19 May 1964	22 May 1964	
I. SAMOS 9	6 Jul 1964	8 Jul 1964	
I. SAMOS 10	14 Aug 1964	23 Aug 1964	
I. SAMOS 11	23 Sep 1964	28 Sep 1964	
I. SAMOS 13	23 Oct 1964	28 Oct 1964	
I. SAMOS 14	4 Dec 1964	5 Dec 1964	
I. SAMOS 15	23 Jan 1965	28 Jan 1965	
I. SAMOS 16	12 Mar 1965	17 Mar 1965	
I. SAMOS 17	28 Apr 1965	3 May 1965	
I. SAMOS 18	27 May 1965	1 Jun 1965	
I. SAMOS 19	25 Jun 1965	30 Jun 1965	
I. SAMOS 21	3 Aug 1965	7 Aug 1965	
I. SAMOS 22	30 Sep 1965	5 Oct 1965	
I. SAMOS 23	8 Nov 1965	9 Nov 1965	
I. SAMOS 24	19 Jan 1966	25 Jan 1966	
I. SAMOS 25	15 Feb 1966	22 Feb 1966	
I. SAMOS 26	18 Mar 1966	24 Mar 1966	
I. SAMOS 27	19 Apr 1966	26 Apr 1966	
I. SAMOS 28	14 May 1966	21 May 1966	
I. SAMOS 29	3 Jun 1966	9 Jun 1966	
I. SAMOS 30	12 Jul 1966	20 Jul 1966	
I. SAMOS 31	16 Aug 1966	24 Aug 1966	
I. SAMOS 32	16 Sep 1966	23 Sep 1966	
T. SAMOS 2	28 Sep 1966	7 Oct 1966	Improved I. SAMOS
I. SAMOS 33	12 Oct 1966	20 Oct 1966	
I. SAMOS 34	2 Nov 1966	10 Nov 1966	
I. SAMOS 35	6 Dec 1966	14 Dec 1966	
T. SAMOS 3	14 Dec 1966	24 Dec 1966	
I. SAMOS 36	2 Feb 1967	12 Feb 1967	
T. SAMOS 4	24 Feb 1967	6 Mar 1967	
I. SAMOS 37	22 May 1967	30 May 1967	

TABLE 2   107

MISSION	LAUNCH	LANDING, RECOVERY, OR RE-ENTRY	NOTES
I. SAMOS 38	4 Jun 1967	12 Jun 1967	
T. SAMOS 6	20 Jun 1967	30 Jun 1967	
T. SAMOS 7	16 Aug 1967	29 Aug 1967	
T. SAMOS 8	19 Sep 1967	30 Sep 1967	
T. SAMOS 9	25 Oct 1967	5 Nov 1967	
T. SAMOS 10	5 Dec 1967	16 Dec 1967	
T. SAMOS 11	18 Jan 1968	4 Feb 1968	
T. SAMOS 12	13 Mar 1968	24 Mar 1968	
T. SAMOS 13	17 Apr 1968	29 Apr 1968	
T. SAMOS 14	5 Jun 1968	17 Jun 1968	
T. SAMOS 15	6 Aug 1968	16 Aug 1968	
T. SAMOS 16	10 Sep 1968	25 Sep 1968	
T. SAMOS 17	6 Nov 1968	20 Nov 1968	
T. SAMOS 18	4 Dec 1968	12 Dec 1968	
T. SAMOS 19	22 Jan 1969	3 Feb 1969	
T. SAMOS 20	4 Mar 1969	18 Mar 1969	
T. SAMOS 21	15 Apr 1969	30 Apr 1969	
T. SAMOS 22	3 Jun 1969	14 Jun 1969	
T. SAMOS 23	23 Aug 1969	7 Sep 1969	
T. SAMOS 24	24 Oct 1969	8 Nov 1969	
T. SAMOS 25	14 Jan 1970	1 Feb 1970	
T. SAMOS 26	15 Apr 1970	6 May 1970	
T. SAMOS 27	25 Jun 1970	6 Jul 1970	
T. SAMOS 28	18 Aug 1970	3 Sep 1970	
T. SAMOS 29	23 Oct 1970	11 Nov 1970	
T. SAMOS 30	21 Jan 1971	9 Feb 1971	Also known as SAMOS 81
T. SAMOS 31	22 Apr 1971	13 May 1971	Also known as SAMOS 82
SAMOS 83	15 Jun 1971	6 Aug 1971	Also known as Big Bird 1
T. SAMOS 32	12 Aug 1971	3 Sep 1971	Also known as SAMOS 84
T. SAMOS 33	23 Oct 1971	17 Nov 1971	
SAMOS 86	20 Jan 1972	29 Feb 1972	Also known as Big Bird 2
T. SAMOS 35	17 Mar 1972	11 Apr 1972	Also known as SAMOS 87
SAMOS 88	7 Jul 1972	13 Sep 1972	Also known as Big Bird 3
T. SAMOS 37	1 Sep 1972	30 Sep 1972	Also known as SAMOS 89
SAMOS 90	10 Oct 1972	8 Jan 1973	Also known as Big Bird 4
T. SAMOS 38	21 Dec 1972	23 Jan 1973	Also known as SAMOS 91
SAMOS 92	9 Mar 1973	19 May 1973	Also known as Big Bird 5
T. SAMOS 39	16 May 1973	13 Jun 1973	Also known as SAMOS 93
SAMOS 94	13 Jul 1973	12 Oct 1973	Also known as Big Bird 6
T. SAMOS 41	27 Sep 1973	28 Oct 1973	Also known as SAMOS 95
SAMOS 96	10 Nov 1973	13 Mar 1974	Also known as Big Bird 7
T. SAMOS 42	13 Feb 1974	17 Mar 1974	Also known as SAMOS 97
SAMOS 98	10 Apr 1974	28 Jul 1974	Also known as Big Bird 8
T. SAMOS 43	6 Jun 1974	23 Jul 1974	Also known as SAMOS 99
T. SAMOS 44	14 Aug 1974	29 Sep 1974	Also known as SAMOS 100
SAMOS 101	29 Oct 1974	19 Mar 1975	Also known as Big Bird 9
T. SAMOS 45	18 Apr 1975	5 Jun 1975	Also known as SAMOS 102
SAMOS 103	8 Jun 1975	5 Nov 1975	Also known as Big Bird 10

MISSION	LAUNCH	LANDING, RECOVERY, OR RE-ENTRY	NOTES
T. SAMOS 46	9 Oct 1975	30 Nov 1975	
T. SAMOS 47	22 Mar 1976	18 May 1976	
T. SAMOS 48	15 Sep 1976	5 Nov 1976	
T. SAMOS 49	13 Mar 1977	26 May 1977	
T. SAMOS 50	23 Sep 1977	8 Dec 1977	
T. SAMOS 51	28 May 1979	26 Aug 1979	

SATURN TESTS (Precursors to Manned Flights)

SA 5	29 Jan 1964	30 Apr 1966	
SA 6	28 May 1964	1 Jun 1964	
SA 7	18 Sep 1964	22 Sep 1964	

SBS: Satellite Business System

SBS 1	15 Nov 1980	
SBS 2	24 Sep 1981	
SBS 3	11 Nov 1982	
SBS 4	31 Aug 1984	

SDS: Satellite Data System

		NOTES
SDS-A	21 Mar 1971	
SDS-B	21 Aug 1973	
SDS 1	10 Mar 1975	
SDS 2	2 Jun 1976	
SDS 3	6 Aug 1976	
SDS 4	25 Feb 1978	
SDS 5	5 Aug 1978	
SDS 6	13 Dec 1980	
SDS 7	24 Apr 1981	
SDS 8	31 Jul 1983	
SDS 9	28 Aug 1984	Also known as USA 4
SDS 10	8 Feb 1985	Also known as USA 9
SDS 11	12 Feb 1987	Also known as USA 21

SECOR: Sequential Collation of Range

		LANDING, RECOVERY, OR RE-ENTRY
SECOR 1	11 Jan 1964	
SECOR 3	9 Mar 1965	
SECOR 2	11 Mar 1965	26 Feb 1968
SECOR 4	3 Apr 1965	
SECOR 5	10 Aug 1965	
SECOR 6	9 Jun 1966	6 Jul 1967
SECOR 7	19 Aug 1966	
SECOR 8	5 Oct 1966	
SECOR 9	29 Jun 1967	
SECOR 13	14 Apr 1969	

SMS: Synchronous Meteorological Satellite

		NOTES
SMS 1	17 May 1974	
SMS 2	6 Feb 1975	
SMS-C	16 Oct 1975	Also known as GOES 1

TABLE 2   109

MISSION	LAUNCH	LANDING, RECOVERY, OR RE-ENTRY	NOTES
SOLRAD (Science, Technology)			
Solrad 1	22 Jun 1960		
Solrad 3	29 Jun 1961		
Solrad 6A	15 Jun 1963	1 Aug 1963	
Solrad 5B	11 Jan 1964		
Solrad 7A	11 Jan 1964		
Solrad 6B	9 Mar 1965		Also known as Ferret 12
Solrad 7B	9 Mar 1965		
Solrad 8	19 Nov 1965		Also known as Explorer 30
Solrad 9	5 Mar 1968		Also known as Explorer 37
Solrad 10	8 Jul 1971	15 Dec 1979	Also known as Explorer 44
Solrad 11A	15 Mar 1976		
Solrad 11B	15 Mar 1976		
SPACENET (Communications)			
Spacenet 1	23 May 1984		
Spacenet 2	10 Nov 1984		
SSU (Reconnaissance and Surveillance)			
SSU 1	8 Jun 1975		
SSU 1	30 Apr 1976		
SSU 2	30 Apr 1976		
SSU 3	30 Apr 1976		
SSU 4	8 Dec 1977		
SSU 5	8 Dec 1977		
SSU 6	8 Dec 1977		
SSU 7	3 Mar 1980		
SSU 8	3 Mar 1980		
SSU 9	18 Jun 1980		
SURCAL: Surveillance Calibration			
SURCAL 1B	12 Dec 1962	18 Jan 1966	
SURCAL 2	12 Dec 1962	5 Feb 1967	
SURCAL 1C	15 Jun 1963	5 Jul 1963	
No Designation	9 Mar 1965	27 Mar 1981	
No Designation	9 Mar 1965		
No Designation	13 Aug 1965		
No Designation	13 Aug 1965		
No Designation	13 Aug 1965		
No Designation	13 Aug 1965		
No Designation	13 Aug 1965		
No Designation	31 May 1967		
No Designation	31 May 1967		
No Designation	31 May 1967		
SYNCOM (Communications)			
Syncom 1	14 Feb 1963		
Syncom 2	26 Jul 1963		
Syncom 3	19 Aug 1964		

MISSION	LAUNCH	LANDING, RECOVERY, OR RE-ENTRY	NOTES
Syncom IV 2	31 Aug 1984		
Syncom IV 1	10 Nov 1984		
Syncom IV 3	13 Apr 1985		
Syncom IV 4	29 Aug 1985		
TELSTAR (Communications)			
Telstar 1	10 Jul 1962		
Telstar 2	7 May 1963		
Telstar 3 A	28 Jul 1983		
Telstar 3 C	1 Sep 1984		
Telstar 3 D	19 Jun 1985		
TETR: Test and Training Satellite			
TETR 1	13 Dec 1967	28 Apr 1968	
TETR 2	8 Nov 1968	19 Sep 1979	
TETR 4	29 Sep 1971	21 Sep 1981	
TIMATION (Navigation, U.S. Navy)			
Timation 1	31 May 1967		
Timation 2	30 Sep 1969	30 Oct 1970	
Timation 3	14 Jul 1974		Also known as NTS 1
TIP: Transit Improvement Program			
TIP 1	2 Sep 1972		
TIP 2	12 Oct 1975		
TIP 3	1 Sep 1976	30 May 1981	
TIROS: Television and Infrared Observation Satellite			
TIROS 1	1 Apr 1960		
TIROS 2	23 Nov 1960		
TIROS 3	12 Jul 1961		
TIROS 4	8 Feb 1962		
TIROS 5	19 Jun 1962		
TIROS 6	18 Sep 1962		
TIROS 7	19 Jun 1963		
TIROS 8	21 Dec 1963		
TIROS 9	22 Jan 1965		
TIROS 10	2 Jul 1965		
TIROS M	23 Jan 1970		
TIROS N	13 Oct 1978		
TRANSIT (Navigation, U.S. Navy)			
Transit 1B	13 Apr 1960	5 Oct 1967	
Transit 2A	22 Jun 1960		
Transit 3B	21 Feb 1961	30 Mar 1961	
Transit 4A	29 Jun 1961		
Transit 4B	15 Nov 1961		
Transit 5A1	19 Dec 1962	25 Sep 1986	
Transit 5A3	16 Jun 1963		
Transit 5BN1	28 Sep 1963		

TABLE 2   111

MISSION	LAUNCH	LANDING, RECOVERY, OR RE-ENTRY	NOTES
Transit 5BN2	5 Dec 1963		
Transit 5C1	3 Jun 1964		
USA (Military)			
USA 1	13 Jun 1984		Also known as Navstar 9
USA 2	25 Jun 1984	18 Oct 1984	Also known as Big Bird 23
USA 3	25 Jun 1984		
USA 4	28 Aug 1984		Also known as SDS 9
USA 5	8 Sep 1984		Also known as Navstar 10
USA 6	4 Dec 1984		Also known as KH 11-6
USA 7	22 Dec 1984		Also known as IMEWS 15
USA 8	25 Jan 1985		
USA 9	8 Feb 1985		Also known as SDS 10
USA 11	3 Oct 1985		Also known as DSCS III 3
USA 12	3 Oct 1985		Also known as DSCS III 4
USA 10	9 Oct 1985		Also known as Navstar 11
USA 13	13 Dec 1985		Also known as ITV 1
USA 14	13 Dec 1985	9 Aug 1987	Also known as ITV 2
USA 15	9 Feb 1986		Also known as NOSS 7
USA 16	9 Feb 1986		
USA 17	9 Feb 1986		
USA 18	9 Feb 1986		
USA 19	5 Sep 1986	28 Sep 1986	SDI test
USA 20	5 Dec 1986		Also known as FLTSATCOM 7
USA 21	12 Feb 1987		Also known as SDS 11
USA 22	15 May 1987		Also known as NOSS 8
USA 23	15 May 1987		
USA 24	15 May 1987		
USA 25	15 May 1987		
USA 26	20 Jun 1987		Also known as DMSP 5D 8
USA 27	26 Oct 1987		Also known as KH 11-7
USA 28	29 Nov 1987		Also known as IMEWS 16
VANGUARD (Science, Technology)			
Vanguard 1	17 Mar 1958		
Vanguard 2	17 Feb 1959		
Vanguard 3	18 Sep 1959		
VELA (Reconnaissance and Surveillance)			
Vela 1	16 Oct 1963		
Vela 2	16 Oct 1963		
Vela 3	17 Jul 1964		
Vela 4	17 Jul 1964		
Vela 5	20 Jul 1965		
Vela 6	20 Jul 1965		
Vela 7	28 Apr 1967		
Vela 8	28 Apr 1967		
Vela 9	23 May 1969		

MISSION	LAUNCH	LANDING, RECOVERY, OR RE-ENTRY	NOTES
Vela 10	23 May 1969		
Vela 11	8 Apr 1970		
Vela 12	8 Apr 1970		
WESTAR (Communications)			
Westar 1	13 Apr 1974		
Westar 2	10 Oct 1974		
Westar 3	10 Aug 1979		
Westar 4	26 Feb 1982		
Westar 5	9 Jun 1982		
Westar 6	3 Feb 1984	16 Nov 1984	

TABLE 3 Unmanned Missions of Other Countries
that Achieved Earth Orbit

MISSION	LAUNCH	LANDING, RECOVERY, OR RE-ENTRY	NOTES
AUSTRALIA			
WRESAT	29 Nov 1967	10 Jan 1968	Weapons Research Establishment Satellite
OSCAR 5	23 Jan 1970		Orbiting Satellite Carrying Amateur Radio (communications)
Aussat 1	27 Aug 1985		Aussat: Communications
Aussat 2	28 Nov 1985		
Aussat 3	16 Sep 1987		
BRAZIL			
Brasilsat 1	8 Feb 1985		Brasilsat: Communications
Brasilsat 2	28 Mar 1986		
BULGARIA			
Bulgaria 1300	7 Aug 1981		Atmospheric studies
CANADA			
ISIS 1	30 Jan 1969		ISIS: International Satellite for Ionospheric Studies
ISIS 2	1 Apr 1971		
Anik A1	10 Nov 1972		Anik: Communications (Anik A1 also known as Telesat 1)
Anik A2	20 Apr 1973		Also known as Telesat 2
Anik A3	7 May 1975		Also known as Telesat 3
CTS 1	17 Jan 1976		Communications Technology Satellite (also known as Hermes)
Anik B	16 Dec 1978		Also known as Telesat 4
Anik D1	26 Aug 1982		Also known as Telesat 5
Anik C3	12 Nov 1982		Also known as Telesat 6
Anik C2	18 Jun 1983		Also known as Telesat 7
Anik D2	9 Nov 1984		Also known as Telesat 8
Anik C1	13 Apr 1985		Also known as Telesat 9
CHINA			
SKW 1	24 Apr 1970		SKW: Shiyan Kexuedi Weixing (space physics)
SKW 2	3 Mar 1971	17 Jun 1979	
SKW 3	26 Jul 1975	14 Sep 1975	
SKW 4	26 Nov 1975	29 Dec 1975	
SKW 5	16 Dec 1975	27 Jan 1976	
SKW 6	30 Aug 1976	25 Nov 1978	
SKW 7	7 Dec 1976	2 Jan 1977	
SKW 8	26 Jan 1978	7 Feb 1978	
SKW 9A	19 Sep 1981	26 Sep 1981	
SKW 9B	19 Sep 1981	6 Oct 1982	
SKW 9C	19 Sep 1981	17 Aug 1982	

MISSION	LAUNCH	LANDING, RECOVERY, OR RE-ENTRY	NOTES
SKW 10	9 Sep 1982	21 Sep 1982	
SKW 11	19 Aug 1983	3 Sep 1983	
STW 1-1	29 Jan 1984		STW: Shiyan Tongxin Weixing (communications)
STW 1-2	8 Apr 1984		
SKW 12	12 Sep 1984	29 Sep 1984	
SKW 13	21 Oct 1985	7 Nov 1985	
STW 2	1 Feb 1986		
SKW 14	6 Oct 1986	23 Oct 1986	
SKW 15	5 Aug 1987	23 Aug 1987	
SKW 16	9 Sep 1987	4 Oct 1987	
CZECHOSLOVAKIA			
Magion	14 Nov 1978	11 Sep 1981	Atmospheric studies
FRANCE			
A1	26 Nov 1965		Rocket motor test
FR 1	6 Dec 1965		Atmospheric studies
D1A Diapason	17 Feb 1966		D1: Science, technology
D1C Diademe-1	8 Feb 1967		
D1D Diademe-2	15 Feb 1967		
Mika	10 Mar 1970	9 Sep 1974	Rocket motor test
Peole	12 Dec 1970	16 Jun 1980	Science, technology
D2A Tournesol	15 Apr 1971	28 Jan 1980	D2: Science, technology
Eole	16 Aug 1971		Meteorological
Oreole 1	27 Dec 1971		Oreole: Atmospheric studies
SRET 1	4 Apr 1972	26 Feb 1974	SRET: Science, technology
Oreole 2	26 Dec 1973		
Symphonie 1	19 Dec 1974		Symphonie: Communications
Starlette	6 Feb 1975		Science, technology
D5A Castor	12 May 1975	18 Feb 1979	D5: Science, technology
D5B Pollux	12 May 1975	5 Aug 1975	
SRET 2	5 Jun 1975		
Symphonie 2	27 Aug 1975		
D2B Aura	27 Sep 1975	30 Sep 1982	
Signe 3	17 Jun 1977	20 Jun 1979	Radiation studies
Oreole 3	21 Sep 1981		
Telecom 1A	4 Aug 1984		Telecom: Communications
Telecom 1B	8 May 1985		
SPOT-1	22 Feb 1986		Satellite Pour Observation de la Terre (earth observation)
GERMANY			
Azur	8 Nov 1969		Radiation studies
Wika	10 Mar 1970	5 Oct 1978	Atmospheric studies (also known as Dial)
Aeros 1	16 Dec 1972	22 Aug 1973	Aeros: Atmospheric studies
Aeros 2	16 Jul 1974	25 Sep 1975	

TABLE 3   115

MISSION	LAUNCH	LANDING, RECOVERY, OR RE-ENTRY	NOTES
Helios 1	10 Dec 1974		Helios: Solar studies
Helios 2	15 Jan 1976		
OSCAR 10	16 Jun 1983		Communications
SPAS 1	22 Jun 1983	22 Jun 1983	Science, technology
IRM	16 Aug 1984		Science, technology
TVSat 1	24 Nov 1987		Communications
INDIA			
Aryabhata	19 Apr 1975		Astronomy
Bhaskara 1	7 Jun 1979		Bhaskara: Science, technology
Rohini 1	18 Jul 1980	20 May 1981	Rohini: Science, technology
Rohini 2	31 May 1981	8 Jun 1981	
Apple 19	Jun 1981		Communications
Bhaskara 2	20 Nov 1981		
Insat 1A	10 Apr 1982		Insat: Communications
Rohini 3	17 Apr 1983		
Insat 1B	31 Aug 1983		
INDONESIA			
Palapa 1	8 Jul 1976		Palapa: Communications
Palapa 2	10 Mar 1977		
Palapa 3	19 Jun 1983		
Palapa 4	4 Feb 1984	16 Nov 1984	
Palapa 5	20 Mar 1987		
ITALY			
San Marco 1	15 Dec 1964	13 Sep 1965	San Marco: Atmospheric studies
San Marco 2	26 Apr 1967	14 Oct 1967	
San Marco 3	24 Apr 1971	29 Nov 1971	
San Marco 4	18 Feb 1974	4 May 1976	
Sirio 1	25 Aug 1977		Communications
JAPAN			
Osumi	11 Feb 1970		Science, technology
MTS 1	16 Feb 1971		MTS: Mu Technology Satellite (also known as Tansei)
Shinsei	28 Sep 1971		Atmospheric studies
Denpa	19 Aug 1972	19 May 1980	Science, technology
MTS 2	16 Feb 1974	22 Jan 1983	
Taiyo	24 Feb 1975	29 Jun 1980	Science
ETS 1	9 Sep 1975		ETS: Engineering Test Satellite (also known as Kiku)
ISS 1	29 Feb 1976		ISS: Ionospheric Sounding Satellite (also known as Ume)
MTS 3	19 Feb 1977		
ETS 2	23 Feb 1977		

MISSION	LAUNCH	LANDING, RECOVERY, OR RE-ENTRY	NOTES
GMS 1	14 Jul 1977		GMS: Geostationary Meteorological Satellite (also known as Himawari)
CS 1	15 Dec 1977		CS: Communication Satellite (also known as Sakura)
Exos 1	4 Feb 1978		Exos: Atmospheric studies (Exos 1 also known as Kyokko)
ISS 2	16 Feb 1978		
BSE	7 Apr 1978		Broadcasting Satellite, Experimental (also known as Yuri)
Exos 2	16 Sep 1978		Also known as Jikiken
ECS 1A	6 Feb 1979		ECS: Experimental Communications Satellite (also known as Ayame); ECS 1A in orbit but not operational
Hakucho	21 Feb 1979	15 Apr 1985	Science
MTS 4	17 Feb 1980	12 May 1983	
ECS 1B	22 Feb 1980		In orbit but not operational
ETS 3	11 Feb 1981		
Astro 1	21 Feb 1981		Astro: Astronomy (Astro 1 also known as Hinotori)
GMS 2	10 Aug 1981		
ETS 4	3 Sep 1982		
CS 2A	4 Feb 1983		
Astro 2	20 Feb 1983	31 Aug 1987	Also known as Tenma
CS 2B	5 Aug 1983		
BS 2A	23 Jan 1984		BS: Broadcasting Satellite (also known as Yuri)
Exos 3	14 Feb 1984	19 Jul 1989	Also known as Ohzora
GMS 3	2 Aug 1984		
BS 2B	12 Feb 1986		
Ajisai	12 Aug 1986		Science, technology
Fuji	12 Aug 1986		Communications (also known as JAS or OSCAR 12)
Mabes	12 Aug 1986		Science, technology
Astro 3	5 Feb 1987		Also known as Ginga
MOS 1	19 Feb 1987		Marine Observation Satellite
ETS 5	27 Aug 1987		
MEXICO			
Morelos 1	17 Jun 1985		Morelos: Communications
Morelos 2	27 Nov 1985		
NETHERLANDS			
ANS	30 Aug 1974	14 Jul 1977	Astronomical Netherlands Satellite
IRAS	26 Jan 1983		Infra Red Astronomical Satellite

TABLE 3   117

MISSION	LAUNCH	LANDING, RECOVERY, OR RE-ENTRY	NOTES
SPAIN			
Intasat	15 Nov 1974		Atmospheric studies
SWEDEN			
Viking	22 Feb 1986		Study of the aurorae
UNITED KINGDOM			
Ariel 1	26 Apr 1962	24 May 1976	Ariel: Science
Ariel 2	27 Mar 1964	18 Nov 1967	
Ariel 3	5 May 1967	14 Dec 1970	
Skynet 1	21 Nov 1969		Skynet: Military communications
X3 Prospero	28 Oct 1971		Science, technology
Ariel 4	11 Dec 1971	12 Dec 1978	
X4 Miranda	9 Mar 1974		Science, technology
Ariel 5	15 Oct 1974	14 Mar 1980	
Skynet 2B	22 Nov 1974		
Ariel 6	2 Jun 1979		
Uosat 1	6 Oct 1981		Uosat: University of Surrey Satellite Communications (Uosat 1 also known as OSCAR 9)
Uosat 2	1 Mar 1984		Also known as OSCAR 11
UKS	16 Aug 1984		UK Subsatellite (science)

TABLE 4 Unmanned International Missions that Achieved Earth Orbit

MISSION	LAUNCH	LANDING, RECOVERY, OR RE-ENTRY	NOTES
AMSAT CORPORATION			
OSCAR 1	12 Dec 1961	31 Jan 1962	OSCAR: Orbiting Satellite Carrying Amateur Radio (communications), satellites of various countries launched by Amsat
OSCAR 2	1 Jun 1962	21 Jun 1962	
OSCAR 3	9 Mar 1965		
OSCAR 4	21 Dec 1965	12 Apr 1976	
OSCAR 5	23 Jan 1970		
OSCAR 6	15 Oct 1972		
OSCAR 7	15 Nov 1974		
OSCAR 8	5 Mar 1978		
OSCAR 9	6 Oct 1981		Also known as Uosat 1
OSCAR 10	16 Jun 1983		
OSCAR 11	1 Mar 1984		Also known as Uosat 2
OSCAR 12	12 Aug 1986		Also known as Fuji
ARAB SATELLITE COMMUNICATIONS ORGANIZATION			
Arabsat 1	8 Feb 1985		
Arabsat 2	18 Jun 1985		
EUROPEAN SPACE AGENCY			
ESRO 2B	17 May 1968	8 May 1971	
ESRO 1A	3 Oct 1968	26 Jun 1970	
HEOS 1	5 Dec 1968	28 Oct 1975	HEOS: Highly Eccentric Orbit Satellite
ESRO 1B	1 Oct 1969	23 Nov 1969	
HEOS 2	31 Jan 1972	2 Aug 1974	
TD 1A	12 Mar 1972	9 Jan 1980	Astronomy
ESRO 4	22 Nov 1972	15 Apr 1974	
COS-B	9 Aug 1975		Astronomy
GEOS 1	20 Apr 1977		GEOS: Geostationary Satellite
ISEE 2	22 Oct 1977	26 Sep 1987	ISEE: International Sun Earth Explorer
Meteosat 1	23 Nov 1977		Meteorology
IUE	26 Jan 1978		International Ultraviolet Explorer
OTS 2	11 May 1978		OTS: Orbital Test Satellite
GEOS 2	14 Jul 1978		
CAT 1	24 Dec 1979		CAT: Capsule Ariane Technologique
CAT 3	19 Jun 1981		
Meteosat 2	19 Jun 1981		
CAT 4	20 Dec 1981		
Marecs 1	20 Dec 1981		Communications
Exosat	26 May 1983	6 May 1986	Astronomy
Eutelsat 1	16 Jun 1983		Communications

TABLE 4   119

MISSION	LAUNCH	LANDING, RECOVERY, OR RE-ENTRY	NOTES
Eutelsat 2	4 Aug 1984		
Marecs B2	10 Nov 1984		
Giotto	2 Jul 1985		Halley's Comet flyby
Eutelsat 4	16 Sep 1987		
INMARSAT			
Marisat 1	19 Feb 1976		Marisat: Maritime communications
Marisat 2	10 Jun 1976		
Marisat 3	14 Oct 1976		
INTELSAT ORGANIZATION			
Intelsat 1	6 Apr 1965		Intelsat: Launched by Intelsat Organization (Intelsat 1 is also known as Early Bird)
Intelsat 2F2	11 Jan 1967		
Intelsat 2F3	22 Mar 1967		
Intelsat 2F4	27 Sep 1967		
Intelsat 3F2	18 Dec 1968		
Intelsat 3F3	5 Feb 1969		
Intelsat 3F4	21 May 1969		
Intelsat 3F5	25 Jul 1969		
Intelsat 3F6	15 Jan 1970		
Intelsat 3F7	23 Apr 1970		
Intelsat 3F8	23 Jul 1970		
Intelsat 4F2	26 Jan 1971		
Intelsat 4F3	19 Dec 1971		
Intelsat 4F4	23 Jan 1972		
Intelsat 4F5	13 Jun 1972		
Intelsat 4F7	23 Aug 1973		
Intelsat 4F8	21 Nov 1974		
Intelsat 4F1	22 May 1975		
Intelsat 4A-F1	26 Sep 1975		
Intelsat 4A-F2	29 Jan 1976		
Intelsat 4A-F4	26 May 1977		
Intelsat 4A-F3	7 Jan 1978		
Intelsat 4A-F6	31 Mar 1978		
Intelsat 5F2	6 Dec 1980		
Intelsat 5F1	23 May 1981		
Intelsat 5F3	15 Dec 1981		
Intelsat 5F4	5 Mar 1982		
Intelsat 5F5	28 Sep 1982		
Intelsat 5F6	19 May 1983		
Intelsat 5F7	19 Oct 1983		
Intelsat 5F8	5 Mar 1984		
Intelsat 5F9	9 Jun 1984	24 Oct 1984	
Intelsat 5F10	22 Mar 1985		
Intelsat 5F11	30 Jun 1985		
Intelsat 5F12	29 Sep 1985		

MISSION	LAUNCH	LANDING, RECOVERY, OR RE-ENTRY	NOTES
INTERKOSMOS			
Interkosmos 1	14 Oct 1969	2 Jan 1970	Solar studies
Interkosmos 2	25 Dec 1969	7 Jun 1970	Atmospheric studies
Interkosmos 3	7 Aug 1970	6 Dec 1970	Solar studies
Interkosmos 4	14 Oct 1970	17 Jan 1971	Solar studies
Interkosmos 5	2 Dec 1971	7 Apr 1972	Solar studies
Interkosmos 6	7 Apr 1972	11 Apr 1972	Science
Interkosmos 7	30 Jun 1972	5 Oct 1972	Solar studies
Interkosmos 8	30 Nov 1972	2 Mar 1973	Solar studies
Interkosmos 9	19 Apr 1973	15 Oct 1973	Solar studies (also known as Copernicus 500)
Interkosmos 10	30 Oct 1973	1 Jul 1977	Atmospheric studies
Interkosmos 11	17 May 1974	6 Sep 1975	Solar studies
Interkosmos 12	31 Oct 1974	11 Jul 1975	Atmospheric studies
Interkosmos 13	27 Mar 1975	2 Sep 1980	Atmospheric studies
Interkosmos 14	11 Dec 1975	27 Feb 1983	Atmospheric studies
Interkosmos 15	19 Jun 1976	18 Nov 1979	Science, technology
Interkosmos 16	27 Jul 1976	10 Jul 1979	Solar studies
Interkosmos 17	24 Sep 1977	8 Nov 1979	Solar studies
Interkosmos 18	24 Oct 1978	17 Mar 1981	Atmospheric studies
Interkosmos 19	27 Feb 1979		Atmospheric studies
Interkosmos 20	1 Nov 1979	5 Mar 1981	Earth studies
Interkosmos 21	6 Feb 1981	7 Jul 1982	Earth studies
Interkosmos 22	7 Aug 1981		Atmospheric studies (also known as Bulgaria 1300)
Interkosmos 23	26 Apr 1985		Also known as Prognoz 10 of U.S.S.R.
NORTH ATLANTIC TREATY ORGANIZATION			
NATO 1	20 Mar 1970		
NATO 2	2 Feb 1971		
NATO 3A	22 Apr 1976		
NATO 3B	28 Jan 1977		
NATO 3C	19 Nov 1978		
NATO 3D	14 Nov 1984		

TABLE 5 Manned Soviet Missions that Achieved Earth Orbit

MISSION	LAUNCH	LANDING	CREW
Vostok 1	12 Apr 1961	12 Apr 1961	Gagarin
Vostok 2	6 Aug 1961	7 Aug 1961	Titov
Vostok 3	11 Aug 1962	15 Aug 1962	Nikolayev
Vostok 4	12 Aug 1962	15 Aug 1962	Popovich
Vostok 5	14 Jun 1963	19 Jun 1963	Bykovsky
Vostok 6	16 Jun 1963	19 Jun 1963	Tereshkova
Voskhod 1	12 Oct 1964	13 Oct 1964	Feoktistov, Komarov, Yegorov
Voskhod 2	18 Mar 1965	19 Mar 1965	Belyayev, Leonov
Soyuz 1	23 Apr 1967	24 Apr 1967	Komarov
Soyuz 3	26 Oct 1968	30 Oct 1968	Beregovoi
Soyuz 4	14 Jan 1969	17 Jan 1969	Shatalov
Soyuz 5	15 Jan 1969	18 Jan 1969	Khrunov, Volynov, Yeliseyev
Soyuz 6	11 Oct 1969	16 Oct 1969	Kubasov, Shonin
Soyuz 7	12 Oct 1969	17 Oct 1969	Filipchenko, Gorbatko, Volkov
Soyuz 8	13 Oct 1969	18 Oct 1969	Shatalov, Yeliseyev
Soyuz 9	2 Jun 1970	19 Jun 1970	Nikolayev, Sevastianov
Soyuz 10	23 Apr 1971	24 Apr 1971	Rukavishnikov, Shatalov, Yeliseyev
Soyuz 11	6 Jun 1971	29 Jun 1971	Dobrovolsky, Patsayev, Volkov
Soyuz 12	27 Sep 1973	29 Sep 1973	Lazarev, Makarov
Soyuz 13	18 Dec 1973	26 Dec 1973	Klimuk, Lebedev
Soyuz 14	3 Jul 1974	19 Jul 1974	Artyukhin, Popovich
Soyuz 15	26 Aug 1974	28 Aug 1974	Demin, Sarafanov
Soyuz 16	2 Dec 1974	8 Dec 1974	Filipchenko, Rukavishnikov
Soyuz 17	10 Jan 1975	9 Feb 1975	Grechko, Gubarev
Soyuz 18	24 May 1975	26 Jul 1975	Klimuk, Sevastianov
Soyuz 19	15 Jul 1975	21 Jul 1975	Kubasov, Leonov
Soyuz 21	6 Jul 1976	24 Aug 1976	Volynov, Zholobov
Soyuz 22	15 Sep 1976	23 Sep 1976	Aksyonov, Bykovsky
Soyuz 23	14 Oct 1976	16 Oct 1976	Rozhdestvensk, Zudov
Soyuz 24	7 Feb 1977	25 Feb 1977	Glazkov, Gorbatko
Soyuz 25	9 Oct 1977	11 Oct 1977	Kovalyonok, Ryumin
Soyuz 26	10 Dec 1977	16 Jan 1978	Grechko, Romanenko. Returned in Soyuz 27.
Soyuz 27	10 Jan 1978	16 Mar 1978	Dzhanibekov, Makarov. Returned in Soyuz 26.
Soyuz 28	2 Mar 1978	10 Mar 1978	Gubarev, Remek
Soyuz 29	15 Jun 1978	3 Sep 1978	Ivanchenkov, Kovalyonok. Returned in Soyuz 31.
Soyuz 30	27 Jun 1978	5 Jul 1978	Hermaszewski, Klimuk
Soyuz 31	26 Aug 1978	2 Nov 1978	Bykovsky, Jaehn. Returned in Soyuz 29.
Soyuz 32	25 Feb 1979	13 Jun 1979	Lyakhov, Ryumin. Returned in Soyuz 34.

MISSION	LAUNCH	LANDING	CREW
Soyuz 33	10 Apr 1979	12 Apr 1979	Ivanov, Rukavishnikov
Soyuz 35	9 Apr 1980	3 Jun 1980	Popov, Ryumin. Returned in Soyuz 37.
Soyuz 36	26 May 1980	30 Jul 1980	Farkas, Kubasov. Returned in Soyuz 35.
Soyuz T2	5 Jun 1980	24 Jun 1980	Aksyonov, Malyshev
Soyuz 37	23 Jul 1980	11 Oct 1980	Gorbatko, Tuan. Returned in Soyuz 36.
Soyuz 38	18 Sep 1980	26 Sep 1980	Mendez, Romanenko
Soyuz T3	27 Nov 1980	10 Dec 1980	Kizim, Makarov, Strekalov
Soyuz T4	12 Mar 1981	26 May 1981	Kovalyonok, Savinykh
Soyuz 39	22 Mar 1981	29 Mar 1981	Dzhanibekov, Gurragcha
Soyuz 40	14 May 1981	22 May 1981	Popov, Prunariu
Soyuz T5	13 May 1982	27 Aug 1982	Berezovoi, Lebedev. Returned in Soyuz T7.
Soyuz T6	24 Jun 1982	2 Jul 1982	Chrétien, Dzhanibekov, Ivanchenkov
Soyuz T7	19 Aug 1982	12 Oct 1982	Popov, Savitskaya, Serebrov. Returned in Soyuz T5.
Soyuz T8	20 Apr 1983	22 Apr 1983	Serebrov, Strekalov, Titov
Soyuz T9	27 Jun 1983	23 Nov 1983	Alexandrov, Lyakhov
Soyuz T10	8 Feb 1984	11 Apr 1984	Atkov, Kizim, Solovyev. Returned in Soyuz T11.
Soyuz T11	3 Apr 1984	2 Oct 1984	Malyshev, Sharma, Strekalov. Returned in Soyuz T10.
Soyuz T12	17 Jul 1984	29 Jul 1984	Dzhanibekov, Savitskaya, Volk
Soyuz T13	6 Jun 1985	25 Sep 1985	Dzhanibekov, Savinykh. Savinykh returned in Soyuz T14.
Soyuz T14	17 Sep 1985	22 Nov 1985	Grechko, Vasyutin, Volkov. Grechko returned in Soyuz T13.
Soyuz T15	13 Mar 1986	16 Jul 1986	Kizim, Solovyev
Soyuz TM2	5 Feb 1987	30 Jul 1987	Laveikin, Romanenko. Romanenko returned in Soyuz TM3.
Soyuz TM3	22 Jul 1987	29 Dec 1987	Alexandrov, Fares, Viktorenko. Viktorenko, Fares returned in Soyuz TM2.
Soyuz TM4	21 Dec 1987	17 Jun 1988	Levchenko, Manarov, Titov. Levchenko returned in Soyuz TM3.

TABLE 6   Manned U.S. Missions that Achieved Earth Orbit

MISSION	LAUNCH	LANDING	CREW
Mercury 6	20 Feb 1962	20 Feb 1962	Glenn
Mercury 7	24 May 1962	24 May 1962	Carpenter
Mercury 8	3 Oct 1962	3 Oct 1962	Schirra
Mercury 9	15 May 1963	16 Mar 1963	Cooper
Gemini 3	23 Mar 1965	23 Mar 1965	Grissom, Young
Gemini 4	3 Jun 1965	7 Jun 1965	McDivitt, White
Gemini 5	21 Aug 1965	29 Aug 1965	Conrad, Cooper
Gemini 7	4 Dec 1965	18 Dec 1965	Borman, Lovell
Gemini 6	15 Dec 1965	12 Dec 1965	Schirra, Stafford
Gemini 8	16 Mar 1966	16 Mar 1966	Armstrong, Scott
Gemini 9	3 Jun 1966	6 Jun 1966	Cernan, Stafford
Gemini 10	18 Jul 1966	21 Jul 1966	Collins, Young
Gemini 11	12 Sep 1966	15 Sep 1966	Conrad, Gordon
Gemini 12	11 Nov 1966	15 Nov 1966	Aldrin, Lovell
Apollo 7	11 Oct 1968	22 Oct 1968	Cunningham, Eisel, Schirra
Apollo 9	3 Mar 1969	13 Mar 1969	McDivitt, Schweickart, Scott
Skylab 2	25 May 1973	22 Jun 1973	Conrad, Kerwin, Weitz
Skylab 3	28 Jul 1973	25 Sep 1973	Bean, Garriott, Lousma
Skylab 4	16 Nov 1973	8 Feb 1974	Carr, Gibson, Pogue
Apollo 18	15 Jul 1975	24 Jul 1975	Brand, Slayton, Stafford
STS 1	12 Apr 1981	14 Apr 1981	Crippen, Young
STS 2	12 Nov 1981	14 Nov 1981	Engle, Truly
STS 3	22 Mar 1982	30 Mar 1982	Fullerton, Lousma
STS 4	27 Jun 1982	4 Jul 1982	Hartsfield, Mattingly
STS 5	11 Nov 1982	16 Nov 1982	Allen, Brand, Lenoir, Overmyer
STS 6	4 Apr 1983	9 Apr 1983	Bobka, Musgrave, Peterson, Weitz
STS 7	18 Jun 1983	24 Jun 1983	Crippen, Fabian, Hauck, Ride, Thagard
STS 8	30 Aug 1983	5 Sep 1983	Bluford, Brandenstein, Gardner, Thornton, Truly
STS 41A	29 Nov 1983	8 Dec 1983	Garriott, Lichtenberg, Merbold, Parker, Shaw, Young
STS 41B	3 Feb 1984	11 Feb 1984	Brand, Gibson, McCandless, McNair, Steward
STS 41C	6 Apr 1984	13 Apr 1984	Crippen, Hart, Hoften, G. Nelson, Scobee
STS 41D	30 Aug 1984	5 Sep 1984	Coats, Hartsfield, Hawley, Mullane, Resnik, C. Walker
STS 41G	5 Oct 1984	13 Oct 1984	Crippen, Garneau, Leestma, McBride, Ride, Scully-Power, Sullivan
STS 51A	8 Nov 1984	16 Nov 1984	Allen, Fisher, Gardner, Hauck, C. Walker

MISSION	LAUNCH	LANDING	CREW
STS 51C	24 Jan 1985	27 Jan 1985	Buchli, Mattingly, Onizuka, Payton, Shriver
STS 51D	12 Apr 1985	19 Apr 1985	Bobka, Garn, Griggs, Hoffman, Seddon, C. Walker, Williams
STS 51B	29 Apr 1985	5 May 1985	Berg, Gregory, Lind, Overmyer, Thagard, Thornton, Wang
STS 51G	17 Jun 1985	24 Jun 1985	Baudry, Brandenstein, Creighton, Fabian, Lucid, Nagel, Al-Saud
STS 51F	29 Jul 1985	6 Aug 1985	Acton, Bartoe, Bridges, England, Fullerton, Henize, Musgrave
STS 51I	27 Aug 1985	3 Sep 1985	Covey, Engle, Fisher, Hoften, Lounge
STS 51J	3 Oct 1985	7 Oct 1985	Bobka, Grabe, Hilmer, Pailes, Stewart
STS 61A	30 Oct 1985	6 Nov 1985	Bluford, Buchli, Dunbar, Furrer, Hartsfield, Messerschmid, Nagel, Ockels
STS 61B	27 Nov 1985	3 Dec 1985	Cleave, O'Connor, Ross, Shaw, Spring, Vela, C. Walker
STS 61C	12 Jan 1986	18 Jan 1986	Bolden, Cenker, Chang-Diaz, Gibson, Hawley, G. Nelson, W. Nelson
STS 26	29 Sep 1988	3 Oct 1988	Covey, Hauck, Hilmers, Lounge, G. Nelson
STS 27	2 Dec 1988	6 Dec 1988	Gardner, Gibson, Mullane, Ross, Shepherd
STS 29	13 Mar 1989	18 Mar 1989	Bagian, Blaha, Buchli, Coats, Springer
STS 30	4 May 1989	8 May 1989	Cleave, Grabe, Lee, Thagard, D. Walker
STS 28	8 Aug 1989	14 Aug 1989	Adamson, Brown, Leestma, Richards, Shaw
STS 34	18 Oct 1989	25 Oct 1989	Baker, Chang-Diaz, Lucid, McCulley, Williams
STS 33	22 Nov 1989	28 Nov 1989	Blaha, Carter, Gregory, Musgrave, Thornton
STS 32	9 Jan 1990	20 Jan 1990	Brandenstein, Dunbar, Ivins, Low, Wetherbee

TABLE 7 Unmanned Lunar Missions

MISSION	COUNTRY	LAUNCH	LANDING	NOTES
Luna 1	U.S.S.R.	2 Jan 1959		Flew past the Moon
Luna 2	U.S.S.R.	12 Sep 1959	15 Sep 1959	Impacted the Moon
Luna 3	U.S.S.R.	4 Oct 1959	29 Apr 1960	Impacted the Moon
Ranger 3	U.S.	26 Jan 1962		Flew past the Moon
Ranger 4	U.S.	23 Apr 1962	26 Apr 1962	Impacted the Moon
Ranger 5	U.S.	18 Oct 1962		Flew past the Moon
Mars 1	U.S.S.R.	1 Nov 1962		Actually launched for Mars but flew past the Moon
Luna 4	U.S.S.R.	2 Apr 1963		Flew past the Moon
Ranger 6	U.S.	30 Jan 1964	2 Feb 1964	Impacted the Moon
Ranger 7	U.S.	28 Jul 1964	31 Jul 1964	Impacted the Moon
Ranger 8	U.S.	17 Feb 1965	20 Feb 1965	Impacted the Moon
Ranger 9	U.S.	21 Mar 1965	24 Mar 1965	Impacted the Moon
Luna 5	U.S.S.R.	9 May 1965	12 May 1965	Impacted the Moon
Luna 6	U.S.S.R.	8 Jun 1965		Flew past the Moon
Zond 3	U.S.S.R.	18 Jul 1965		Flew past the Moon
Luna 7	U.S.S.R.	4 Oct 1965	7 Oct 1965	Impacted the Moon
Luna 8	U.S.S.R.	3 Dec 1965	6 Dec 1965	Impacted the Moon
Luna 9	U.S.S.R.	31 Jan 1966	3 Feb 1965	Soft-landed on the Moon
Luna 10	U.S.S.R.	31 Mar 1966		In lunar orbit; may have impacted the Moon by now
Surveyor 1	U.S.	30 May 1966	2 Jun 1966	Soft-landed on the Moon
Orbiter 1	U.S.	10 Aug 1966	29 Oct 1966	Impacted the Moon
Luna 11	U.S.S.R.	24 Aug 1966		In lunar orbit; may have impacted the Moon by now
Surveyor 2	U.S.	20 Sep 1966	23 Sep 1966	Impacted the Moon
Luna 12	U.S.S.R.	22 Oct 1966		In lunar orbit; may have impacted the Moon by now
Orbiter 2	U.S.	6 Nov 1966	11 Oct 1967	Impacted the Moon
Luna 13	U.S.S.R.	21 Dec 1966	24 Dec 1966	Soft-landed on the Moon
Orbiter 3	U.S.	4 Feb 1967	9 Oct 1967	Impacted the Moon
Surveyor 3	U.S.	17 Apr 1967	20 Apr 1967	Soft-landed on the Moon
Orbiter 4	U.S.	4 May 1967	6 Oct 1967	Impacted the Moon
Surveyor 4	U.S.	14 Jul 1967	17 Jul 1967	Impacted the Moon
Orbiter 5	U.S.	1 Aug 1967	31 Jan 1968	Impacted the Moon
Surveyor 5	U.S.	8 Sep 1967	11 Sep 1967	Soft-landed on the Moon
Surveyor 6	U.S.	7 Nov 1967	10 Nov 1967	Soft-landed on the Moon
Surveyor 7	U.S.	7 Jan 1968	10 Jan 1968	Soft-landed on the Moon
Luna 14	U.S.S.R.	7 Apr 1968		In lunar orbit; may have impacted the Moon by now
Luna 15	U.S.S.R.	13 Jul 1969	21 Jul 1969	Impacted the Moon

MISSION	COUNTRY	LAUNCH	LANDING	NOTES
Luna 16	U.S.S.R.	12 Sep 1970	24 Sep 1970	Soft-landed on the Moon; returned to Earth with lunar soil sample
Luna 17	U.S.S.R.	10 Nov 1970	17 Nov 1970	Soft-landed on the Moon
Luna 18	U.S.S.R.	2 Sep 1971	11 Sep 1971	Impacted the Moon
Luna 19	U.S.S.R.	28 Sep 1971		In lunar orbit; may have impacted the Moon by now
Luna 20	U.S.S.R.	14 Feb 1972	25 Feb 1972	Soft-landed on the Moon; returned to Earth with lunar soil sample
Luna 21	U.S.S.R.	8 Jan 1973	16 Jan 1973	Soft-landed on the Moon
Luna 22	U.S.S.R.	29 May 1974		In lunar orbit; may have impacted the Moon by now
Luna 23	U.S.S.R.	28 Oct 1974	6 Nov 1974	Soft-landed on the Moon
Luna 24	U.S.S.R.	9 Aug 1976	18 Aug 1976	Soft-landed on the Moon; returned to Earth with lunar soil sample

TABLE 8 Manned Lunar Missions

MISSION	COUNTRY	LAUNCH	RETURN	CREW & NOTES
Apollo 8	U.S.	21 Dec 1968	27 Dec 1968	Anders, Borman, Lovell. Did not land on the Moon.
Apollo 10	U.S.	18 May 1969	26 May 1969	Cernan, Stafford, Young. Did not land on the Moon.
Apollo 11[a]	U.S.	16 Jul 1969	24 Jul 1969	*Aldrin (2)*, *Armstrong (1)*, Collins
Apollo 12[b]	U.S.	14 Nov 1969	24 Nov 1969	*Bean (4)*, *Conrad (3)*, Gordon
Apollo 13	U.S.	11 Apr 1970	17 Apr 1970	Haise, Lovell, Swigert. Could not land on the Moon
Apollo 14[c]	U.S.	31 Jan 1971	9 Feb 1971	*Mitchell (6)*, Roosa, *Shepard (5)*
Apollo 15[d]	U.S.	26 Jul 1971	7 Aug 1971	*Irwin (8)*, *Scott (7)*, Worden
Apollo 16[e]	U.S.	16 Apr 1972	27 Apr 1972	*Duke (10)*, Mattingly, *Young (9)*
Apollo 17[f]	U.S.	7 Dec 1972	19 Dec 1972	*Cernan (11)*, Evans, *Schmitt (12)*

Names in italics landed on the Moon. Numbers in parentheses indicate the numerical ranking of those who have set foot on the Moon.

[a] **Lunar landing 20 Jul 1969; lift-off 21 Jul 1969.**

[b] **Lunar landing 19 Nov 1969; lift-off 20 Nov 1969.**

[c] **Lunar landing 5 Feb 1971; lift-off 6 Feb 1971.**

[d] **Lunar landing 30 Jul 1971; lift-off 2 Aug 1971.**

[e] **Lunar landing 20 Apr 1972; lift-off 23 Apr 1972.**

[f] **Lunar landing 11 Dec 1972; lift-off 14 Dec 1972.**

TABLE 9   Space Explorers, Their Missions, and Mission Durations

SPACE EXPLORER	COUNTRY	MISSION	YEAR	DURATION DAYS	HRS.
Loren Acton	U.S.	STS 51F	1985	7	23
James Adamson	U.S.	STS 28	1989	6	
Vladimir Aksyonov	U.S.S.R.	Soyuz 22	1976	7	22
		Soyuz T2	1980	3	22
Salman Al-Saud	Saudi Arabia	STS 51G	1985	7	2
Edwin Aldrin	U.S.	Gemini 12	1966	3	23
		Apollo 11	1969	8	3
Alexander Alexandrov	U.S.S.R.	Soyuz T9	1983	149	10
		Soyuz TM3	1987	160	7
		Soyuz TM5	1988	92	
Joseph Allen	U.S.	STS 5	1982	5	2
		STS 51A	1984	8	
William Anders	U.S.	Apollo 8	1986	6	3
Neil Armstrong	U.S.	Gemini 8	1966		11
		Apollo 11	1969	8	3
Yuri Artyukhin	U.S.S.R.	Soyuz 14	1974	15	18
Oleg Atkov	U.S.S.R.	Soyuz T10	1984	236	23
James Bagian	U.S.	STS 29	1989	5	
Ellen Baker	U.S.	STS 34	1989	7	
John-David Bartoe	U.S.	STS 51F	1985	7	23
Patrick Baudry	France	STS 51G	1985	7	2
Alan Bean	U.S.	Apollo 12	1969	10	5
		Skylab 3	1973	59	11
Pavel Belyayev	U.S.S.R.	Voskhod 2	1965	1	2
Georgi Beregovoi	U.S.S.R.	Soyuz 3	1968	3	23
Anatoly Berezovoi	U.S.S.R.	Soyuz T5	1982	211	9
John Blaha	U.S.	STS 29	1989	5	
		STS 33	1989	6	
Guion Bluford	U.S.	STS 8	1983	6	1
		STS 61A	1985	7	1
Karol Bobka	U.S.	STS 6	1983	5	
		STS 51D	1985	7	
		STS 51J	1985	4	1
Charles Bolden	U.S.	STS 61C	1986	6	2
Frank Borman	U.S.	Gemini 7	1965	13	19
		Apollo 8	1968	6	3
Vance Brand	U.S.	Apollo 18	1975	9	1
		STS 5	1982	5	2
		STS 41B	1984	7	23
Daniel Brandenstein	U.S.	STS 8	1983	6	1
		STS 51G	1985	7	2
		STS 32	1990	11	

Note: Only those space explorers who have completed at least one Earth orbit are listed. Some explorers exchanged spacecraft in orbit, returning to Earth in spacecraft different from those in which they were launched. U.S. explorers are listed up to 1990; Soviet explorers are listed up to 1988. Mission durations have been rounded off to the nearest day and hour.

TABLE 9   129

SPACE EXPLORER	COUNTRY	MISSION	YEAR	DURATION DAYS	HRS.
Roy Bridges	U.S.	STS 51F	1985	7	23
Mark Brown	U.S.	STS 28	1989	6	
James Buchli	U.S.	STS 51C	1985	3	2
		STS 61A	1985	7	1
		STS 29	1989	5	
Valery Bykovsky	U.S.S.R.	Vostok 5	1963	4	23
		Soyuz 22	1976	7	22
		Soyuz 31	1978	7	21
Scott Carpenter	U.S.	Mercury-Aurora 7	1962		5
Gerald Carr	U.S.	Skylab 4	1973	84	1
Manley Carter	U.S.	STS 33	1989	6	
Robert Cenker	U.S.	STS 61C	1986	6	2
Eugene Cernan	U.S.	Gemini 9	1966	3	
		Apollo 10	1969	8	
		Apollo 17	1972	12	14
Franklin Chang-Diaz	U.S.	STS 61C	1986	6	2
		STS 34	1989	7	
Jean-Loup Chrétien	France	Soyuz T6	1982	7	22
Mary Cleave	U.S.	STS 61B	1985	6	21
		STS 30	1989	4	
Michael Coats	U.S.	STS 41D	1984	6	1
		STS 29	1989	5	
Michael Collins	U.S.	Gemini 10	1966	2	23
		Apollo 11	1969	8	3
Charles Conrad	U.S.	Gemini 5	1965	7	22
		Gemini 11	1966	2	23
		Apollo 12	1969	10	5
		Skylab 2	1973	28	
Gordon Cooper	U.S.	Mercury-Faith 7	1963	1	10
		Gemini 5	1965	7	23
Richard Covey	U.S.	STS 51I	1985	7	2
		STS 26	1988	4	
John Creighton	U.S.	STS 51G	1985	7	2
Robert Crippen	U.S.	STS 1	1981	2	6
		STS 7	1983	6	2
		STS 41C	1984	7	23
		STS 41G	1984	8	5
Walter Cunningham	U.S.	Apollo 7	1968	10	20
Lev Demin	U.S.S.R.	Soyuz 15	1974	2	
Georgi Dobrovolsky	U.S.S.R.	Soyuz 11	1971	23	18
Charles Duke	U.S.	Apollo 16	1972	11	2
Bonnie Dunbar	U.S.	STS 61A	1985	7	1
		STS 32	1990	11	
Vladimir Dzhanibekov	U.S.S.R.	Soyuz 27	1978	(Total time stated	
		Soyuz 39	1981	for 5 missions	
		Soyuz T6	1982	combined is	
		Soyuz T12	1984	145 days, 16 hrs.)	
		Soyuz T13	1985		

SPACE EXPLORER	COUNTRY	MISSION	YEAR	DURATION DAYS	HRS.
Donn Eisele	U.S.	Apollo 7	1968	10	20
Anthony England	U.S.	STS 51F	1985	7	23
Joe Engle	U.S.	STS 2	1981	2	6
		STS 51I	1985	7	2
Ronald Evans	U.S.	Apollo 17	1972	12	14
John Fabian	U.S.	STS 7	1983	6	2
		STS 51G	1985	7	2
Mohammed Faris	Syria	Soyuz TM3	1987	7	23
Bertalan Farkas	Hungary	Soyuz 36	1980	7	21
Konstantin Feoktistov	U.S.S.R.	Voskhod 1	1965	1	
Anatoli Filipchenko	U.S.S.R.	Soyuz 7	1969	4	23
		Soyuz 16	1974	5	22
Anna Fisher	U.S.	STS 51A	1984	8	
William Fisher	U.S.	STS 51I	1985	7	2
Charles Fullerton	U.S.	STS 3	1982	8	
		STS 51F	1985	7	23
Rheinhard Furrer	West Germany	STS 61A	1985	7	1
Yuri Gagarin	U.S.S.R.	Vostok 1	1961		2
Dale Gardner	U.S.	STS 8	1983	6	1
		STS 51A	1984	8	
Guy Gardner	U.S.	STS 27	1988	4	
Jake Garn	U.S.	STS 51D	1985	7	
Marc Garneau	Canada	STS 41G	1984	8	5
Owen Garriott	U.S.	Skylab 3	1973	59	11
		STS 9	1983	10	8
Edward Gibson	U.S.	STS 41B	1984	7	23
		STS 61C	1986	6	2
Robert Gibson	U.S.	Skylab 4	1973	84	1
		STS 27	1988	4	
Yuri Glazkov	U.S.S.R.	Soyuz 24	1977	17	17
John Glenn	U.S.	Mercury-Friendship 7	1962		5
Viktor Gorbatko	U.S.S.R.	Soyuz 7	1969	4	23
		Soyuz 24	1977	17	17
		Soyuz 37	1980	7	21
Richard Gordon	U.S.	Gemini 11	1966	2	23
		Apollo 12	1969	10	5
Ronald Grabe	U.S.	STS 51J	1985	4	1
		STS 30	1989	4	
Georgi Grechko	U.S.S.R.	Soyuz 17	1975	29	13
		Soyuz 26	1977	96	10
		Soyuz T14	1985	8	21
Frederick Gregory	U.S.	STS 51B	1985	7	
		STS 33	1989	6	
David Griggs	U.S.	STS 51D	1985	7	
Virgil Grissom	U.S.	Mercury 4	1961		
		Gemini 3	1965		5
Alexei Gubarev	U.S.S.R.	Soyuz 17	1975	29	13
		Soyuz 28	1978	7	22

TABLE 9   131

SPACE EXPLORER	COUNTRY	MISSION	YEAR	DURATION DAYS	HRS.
Jugderdemidyn Gurragcha	Mongolia	Soyuz 39	1981	7	21
Fred Haise	U.S.	Apollo 13	1970	5	23
Terry Hart	U.S.	STS 41C	1984	7	
Henry Hartsfield	U.S.	STS 4	1982	7	1
		STS 41D	1984	6	1
		STS 61A	1985	7	1
Frederick Hauck	U.S.	STS 7	1983	6	2
		STS 51A	1984	8	
		STS 26	1988	4	
Steven Hawley	U.S.	STS 41D	1984	6	1
		STS 61C	1986	6	2
Karl Henize	U.S.	STS 51F	1985	7	23
Miroslaw Hermaszewski	Poland	Soyuz 30	1978	7	22
David Hilmers	U.S.	STS 51J	1985	4	1
		STS 26	1988	4	
Jeffrey Hoffman	U.S.	STS 51D	1985	7	
James van Hoften	U.S.	STS 41C	1984	7	
		STS 51I	1985	7	2
James Irwin	U.S.	Apollo 15	1971	12	7
Alexander Ivanchenkov	U.S.S.R.	Soyuz 29	1978	139	15
		Soyuz T6	1982	7	22
Georgi Ivanov	Bulgaria	Soyuz 33	1979	1	23
Marsha Ivins	U.S.	STS 32	1990	11	
Sigmund Jaehn	East Germany	Soyuz 31	1978	7	21
Joseph Kerwin	U.S.	Skylab 2	1973	28	1
Yevgeny Khrunov	U.S.S.R.	Soyuz 5	1969	2	
Leonid Kizim	U.S.S.R.	Soyuz T3	1980	12	19
		Soyuz T10	1984	236	23
		Soyuz T15	1986	125	
Pyotr Klimuk	U.S.S.R.	Soyuz 13	1973	7	21
		Soyuz 18	1975	62	23
		Soyuz 30	1978	7	22
Vladimir Komarov	U.S.S.R.	Voskhod 1	1964	1	
		Soyuz 1	1967	1	3
Vladimir Kovalyonok	U.S.S.R.	Soyuz 25	1977	2	1
		Soyuz 29	1978	139	15
		Soyuz T4	1981	74	19
Valery Kubasov	U.S.S.R.	Soyuz 6	1969	4	23
		Soyuz 19	1975	5	23
		Soyuz 36	1980	7	21
Alexander Laveikin	U.S.S.R.	Soyuz TM2	1987	174	3
Vasily Lazarev	U.S.S.R.	Soyuz 12	1973	1	23
Valentin Lebedev	U.S.S.R.	Soyuz 13	1973	7	21
		Soyuz T5	1982	211	9
Mark Lee	U.S.	STS 30	1989	4	
David Leestma	U.S.	STS 41G	1984	8	5
		STS 28	1989	6	

SPACE EXPLORER	COUNTRY	MISSION	YEAR	DURATION DAYS	HRS.
William Lenoir	U.S.	STS 5	1982	5	2
Alexei Leonov	U.S.S.R.	Voskhod 2	1965	1	2
		Soyuz 19	1975	5	23
Anatoly Levchenko	U.S.S.R.	Soyuz TM4	1987	7	22
Byron Lichtenberg	U.S.	STS 9	1983	10	8
Don Lind	U.S.	STS 51B	1985	7	
John Lounge	U.S.	STS 51I	1985	7	2
		STS 26	1988	4	
Jack Lousma	U.S.	Skylab 3	1973	59	11
		STS 3	1982	8	
James Lovell	U.S.	Gemini 7	1965	13	19
		Gemini 12	1966	3	23
		Apollo 8	1968	6	3
		Apollo 13	1970	5	23
David Low	U.S.	STS 32	1990	11	
Shannon Lucid	U.S.	STS 51G	1985	7	2
		STS 34	1989	7	
Vladimir Lyakhov	U.S.S.R.	Soyuz 32	1979	175	1
		Soyuz T9	1983	149	10
Jon McBride	U.S.	STS 41G	1984	8	5
Bruce McCandless	U.S.	STS 41B	1984	7	23
Michael McCulley	U.S.	STS 34	1989	7	
James McDivitt	U.S.	Gemini 4	1965	4	2
		Apollo 9	1969	10	1
Ronald McNair	U.S.	STS 41B	1984	7	23
Oleg Makarov	U.S.S.R.	Soyuz 12	1973	1	23
		Soyuz 27	1978	5	23
		Soyuz T3	1980	12	19
Yuri Malyshev	U.S.S.R.	Soyuz T2	1980	3	22
		Soyuz T11	1984	7	22
Musa Manarov	U.S.S.R.	Soyuz TM4	1987	178	
Thomas Mattingly	U.S.	Apollo 16	1972	11	2
		STS 4	1982	7	1
		STS 51C	1985	3	2
Arnaldo Tamayo Mendez	Cuba	Soyuz 38	1980	7	21
Ulf Merbold	West Germany	STS 9	1983	10	8
Ernst Messerschmid	West Germany	STS 61A	1985	7	1
Edgar Mitchell	U.S.	Apollo 14	1971	9	
Richard Mullane	U.S.	STS 41D	1984	6	1
		STS 27	1988	4	
Story Musgrave	U.S.	STS 6	1983	5	
		STS 51F	1985	7	23
		STS 33	1989	6	
Steven Nagel	U.S.	STS 51G	1985	7	2
		STS 61A	1985	7	1
Bill Nelson	U.S.	STS 61C	1986	6	2
George Nelson	U.S.	STS 41C	1984	7	
		STS 61C	1986	6	2
		STS 26	1988	4	

TABLE 9   133

SPACE EXPLORER	COUNTRY	MISSION	YEAR	DURATION DAYS	HRS.
Andrian Nikolayev	U.S.S.R.	Vostok 3	1962	3	22
		Soyuz 9	1970	17	17
Wubbo Ockels	Netherlands	STS 61A	1985	7	1
Bryan O'Connor	U.S.	STS 61B	1985	6	21
Ellison Onizuka	U.S.	STS 51C	1985	3	2
Robert Overmyer	U.S.	STS 5	1982	5	2
		STS 51B	1985	7	
William Pailes	U.S.	STS 51J	1985	4	1
Robert Parker	U.S.	STS 9	1983	10	8
Viktor Patsayev	U.S.S.R.	Soyuz 11	1971	23	18
Gary Payton	U.S.	STS 51C	1985	3	2
Donald Peterson	U.S.	STS 6	1983	5	
William Pogue	U.S.	Skylab 4	1973	84	1
Leonid Popov	U.S.S.R.	Soyuz 35	1980	184	20
		Soyuz 40	1981	7	21
		Soyuz T7	1982	7	22
Pavel Popovich	U.S.S.R.	Vostok 4	1962	2	23
		Soyuz 14	1974	15	18
Dumitru Prunariu	Romania	Soyuz 40	1981	7	21
Vladimir Remek	Czechoslovakia	Soyuz 28	1978	7	22
Judith Resnik	U.S.	STS 41D	1984	6	1
Richard Richards	U.S.	STS 28	1989	6	
Sally Ride	U.S.	STS 7	1983	6	2
		STS 41G	1984	8	5
Yuri Romanenko	U.S.S.R.	Soyuz 26	1977	96	10
		Soyuz 38	1980	7	21
		Soyuz TM2	1987	326	12
Stuart Roosa	U.S.	Apollo 14	1971	9	
Jerry Ross	U.S.	STS 61B	1985	6	21
		STS 27	1988	4	
Valery Rozhdestvensky	U.S.S.R.	Soyuz 23	1976	2	
Nikolai Rukavishnikov	U.S.S.R.	Soyuz 10	1971	2	
		Soyuz 16	1974	5	22
		Soyuz 33	1979	1	23
Valery Ryumin	U.S.S.R.	Soyuz 25	1977	2	1
		Soyuz 32	1979	175	1
		Soyuz 35	1980	184	20
Gennady Sarafanov	U.S.S.R.	Soyuz 15	1974	2	
Viktor Savinykh	U.S.S.R.	Soyuz T4	1981	74	19
		Soyuz T13	1985	168	4
		Soyuz TM5	1988	92	
Svetlana Savitskaya	U.S.S.R.	Soyuz T7	1982	7	22
		Soyuz T12	1984	11	19
Walter Schirra	U.S.	Mercury-Sigma 7	1962		9
		Gemini 6	1965	1	2
		Apollo 7	1968	10	20
Harrison Schmitt	U.S.	Apollo 17	1972	12	14
Russell Schweickart	U.S.	Apollo 9	1969	10	1
Francis Scobee	U.S.	STS 41C	1984	7	

SPACE EXPLORER	COUNTRY	MISSION	YEAR	DURATION DAYS	HRS.
David Scott	U.S.	Gemini 8	1966	10	41
		Apollo 9	1969	10	1
		Apollo 15	1971	12	7
Paul Scully-Power	U.S.	STS 41G	1984	8	5
Margaret Rhea Seddon	U.S.	STS 51D	1985	7	
Alexander Serebrov	U.S.S.R.	Soyuz T7	1982	7	22
		Soyuz T8	1983	2	
Vitaly Sevastianov	U.S.S.R.	Soyuz 9	1970	17	17
		Soyuz 18	1975	62	23
Rakesh Sharma	India	Soyuz T11	1984	7	22
Vladimir Shatalov	U.S.S.R.	Soyuz 4	1969	2	23
		Soyuz 8	1969	4	23
		Soyuz 10	1971	2	
Brewster Shaw	U.S.	STS 9	1983	10	8
		STS 61B	1985	6	21
		STS 28	1989	6	
Alan Shepard	U.S.	Apollo 14	1971	9	
William Shepherd	U.S.	STS 27	1988	4	
Georgi Shonin	U.S.S.R.	Soyuz 6	1969	4	23
Loren Shriver	U.S.	STS 51C	1985	3	2
Donald Slayton	U.S.	Apollo 18	1975	9	1
Vladimir Solovyev	U.S.S.R.	Soyuz T10	1984	236	23
		Soyuz T15	1986	125	
		Soyuz TM5	1988	92	
Sherwood Spring	U.S.	STS 61B	1985	6	21
Robert Springer	U.S.	STS 29	1989	5	
Thomas Stafford	U.S.	Gemini 6	1965	1	2
		Gemini 9	1966	3	
		Apollo 10	1969	8	
		Apollo 18	1975	9	1
Robert Stewart	U.S.	STS 41B	1984	7	23
		STS 51J	1985	4	1
Gennady Strekalov	U.S.S.R.	Soyuz T3	1980	12	19
		Soyuz T8	1983	2	
		Soyuz T11	1984	7	22
Kathryn Sullivan	U.S.	STS 41G	1984	8	5
John Swigert	U.S.	Apollo 13	1970	5	23
Valentina Tereshkova	U.S.S.R.	Vostok 6	1963	2	23
Norman Thagard	U.S.	STS 7	1983	6	2
		STS 51B	1985	7	
		STS 30	1989	4	
Kathryn Thornton	U.S.	STS 33	1989	6	
William Thornton	U.S.	STS 8	1983	6	1
		STS 51B	1985	7	
Gherman Titov	U.S.S.R.	Vostok 2	1961	1	1
Vladimir Titov	U.S.S.R.	Soyuz T8	1983	2	
		Soyuz TM4	1987	178	
Richard Truly	U.S.	STS 2	1981	2	6
		STS 8	1983	6	1

TABLE 9   135

SPACE EXPLORER	COUNTRY	MISSION	YEAR	DURATION DAYS	HRS.
Pham Tuan	Vietnam	Soyuz 37	1980	7	21
Lodewijk van den Berg	U.S.	STS 51B	1985	7	
Vladimir Vasyutin	U.S.S.R.	Soyuz T14	1985	64	22
Rodolfo Neri Vela	Mexico	STS 61B	1985	6	21
Alexander Viktorenko	U.S.S.R.	Soyuz TM3	1987	7	23
Igor Volk	U.S.S.R.	Soyuz T12	1984	11	19
Alexander Volkov	U.S.S.R.	Soyuz T14	1985	64	22
Vladislav Volkov	U.S.S.R.	Soyuz 7	1969	4	23
		Soyuz 11	1971	23	18
Boris Volynov	U.S.S.R.	Soyuz 5	1969	3	1
		Soyuz 21	1976	49	6
Charles Walker	U.S.	STS 41D	1984	6	1
		STS 51D	1985	7	
		STS 61B	1985	6	21
David Walker	U.S.	STS 51A	1984	8	
		STS 30	1989	4	
Taylor Wang	U.S.	STS 51B	1985	7	
Paul Weitz	U.S.	Skylab 2	1973	28	1
		STS 6	1983	5	
James Wetherbee	U.S.	STS 32	1990	11	
Edward White	U.S.	Gemini 4	1965	4	2
Donald Williams	U.S.	STS 51D	1985	7	
		STS 34	1989	7	
Alfred Worden	U.S.	Apollo 15	1971	12	7
Boris Yegorov	U.S.S.R.	Voskhod 1	1964	1	
Alexei Yeliseyev	U.S.S.R.	Soyuz 5	1969	2	
		Soyuz 8	1969	4	23
		Soyuz 10	1971	2	
John Young	U.S.	Gemini 3	1965		5
		Gemini 10	1966	2	23
		Apollo 10	1969	8	
		Apollo 16	1972	11	2
		STS 1	1981	2	6
		STS 9	1983	10	8
Vitaly Zholobov	U.S.S.R.	Soyuz 21	1976	49	6
Vyacheslav Zudov	U.S.S.R.	Soyuz 23	1976	2	

TABLE 10 Interplanetary Missions

MISSION	COUNTRY	LAUNCH	LANDING	NOTES
Pioneer 4	U.S.	3 Mar 1959		Solar probe
Pioneer 5	U.S.	11 Mar 1960		Solar probe
Venera 1	U.S.S.R.	12 Feb 1961		Venus flyby
Mariner 2	U.S.	27 Aug 1962		Venus flyby
Zond 1	U.S.S.R.	2 Apr 1964		Venus flyby
Mariner 3	U.S.	5 Nov 1964		Mars flyby, now in solar orbit
Mariner 4	U.S.	28 Nov 1964		Mars flyby
Zond 2	U.S.S.R.	30 Nov 1964		Mars flyby
Venera 2	U.S.S.R.	12 Nov 1965		Venus flyby
Venera 3	U.S.S.R.	16 Nov 1965	1 Mar 1966	Impacted on Venus
Pioneer 6	U.S.	16 Dec 1965		Solar probe
Pioneer 7	U.S.	17 Aug 1966		Solar probe
Venera 4	U.S.S.R.	12 Jun 1967	18 Oct 1967	Impacted on Venus
Mariner 5	U.S.	14 Jun 1967		Venus flyby
Pioneer 8	U.S.	13 Dec 1967		Solar probe
Zond 5	U.S.S.R.	15 Sep 1968	21 Sep 1968	Destination not certain
Pioneer 9	U.S.	8 Nov 1968		Solar probe
Zond 6	U.S.S.R.	10 Nov 1968	17 Nov 1968	Destination not certain
Venera 5	U.S.S.R.	5 Jan 1969	16 May 1969	Impacted on Venus
Venera 6	U.S.S.R.	10 Jan 1969	17 May 1969	Impacted on Venus
Mariner 6	U.S.	24 Feb 1969		Mars flyby
Mariner 7	U.S.	27 Mar 1969		Mars flyby
Zond 7	U.S.S.R.	8 Aug 1969	14 Aug 1969	Destination not certain
Venera 7	U.S.S.R.	17 Aug 1970	15 Dec 1970	Soft-landed on Venus
Zond 8	U.S.S.R.	20 Oct 1970	27 Oct 1970	Destination not certain
Mars 2	U.S.S.R.	19 May 1971	27 Nov 1971	Mars orbit, then impact
Mars 3	U.S.S.R.	28 May 1971	2 Dec 1971	Soft-landed on Mars
Mariner 9	U.S.	30 May 1971		Mars orbit
Pioneer 10	U.S.	2 Mar 1972		Jupiter flyby
Venera 8	U.S.S.R.	27 Mar 1972	22 Jul 1972	Soft-landed on Venus
Pioneer 11	U.S.	5 Apr 1973		Jupiter, Saturn flyby
Mars 4	U.S.S.R.	21 Jul 1973		Mars flyby
Mars 5	U.S.S.R.	25 Jul 1973		Mars orbit
Mars 6	U.S.S.R.	5 Aug 1973	12 Mar 1975	Soft-landed on Mars
Mars 7	U.S.S.R.	9 Aug 1973		Mars flyby
Mariner 10	U.S.	3 Nov 1973		Venus, Mercury flyby
Venera 9	U.S.S.R.	8 Jun 1975	22 Oct 1975	Soft-landed on Venus
Venera 10	U.S.S.R.	14 Jun 1975	25 Oct 1975	Soft-landed on Venus
Viking 1	U.S.	20 Aug 1975	20 Jul 1976	Soft-landed on Mars
Viking 2	U.S.	9 Sep 1975	3 Sep 1976	Soft-landed on Mars
Voyager 2	U.S.	20 Aug 1977		Jupiter, Saturn, Uranus, Neptune flyby
Voyager 1	U.S.	5 Sep 1977		Jupiter, Saturn flyby
Venus 1	U.S.	20 May 1978		Venus orbit
Venus 2	U.S.	8 Aug 1978	9 Dec 1978	Four atmospheric probes launched from Venus orbit
Venera 11	U.S.S.R.	9 Sep 1978	25 Dec 1978	Soft-landed on Venus

TABLE 10   137

MISSION	COUNTRY	LAUNCH	LANDING	NOTES
Venera 12	U.S.S.R.	14 Sep 1978	21 Dec 1978	Soft-landed on Venus
Venera 13	U.S.S.R.	30 Oct 1981	1 Mar 1982	Soft-landed on Venus
Venera 14	U.S.S.R.	4 Nov 1981	5 Mar 1982	Soft-landed on Venus
Venera 15	U.S.S.R.	2 Jun 1983		Venus orbit
Venera 16	U.S.S.R.	7 Jun 1983		Venus orbit
Vega 1	U.S.S.R.	15 Dec 1984		Ejected Venus lander 10 Jun 1985, then went on for Halley's Comet flyby
Vega 2	U.S.S.R.	21 Dec 1984		Ejected Venus lander 14 Jun 1985, then went on for Halley's Comet flyby
Sakigake	Japan	7 Jan 1985		Halley's Comet flyby
Giotto	ESA	2 Jul 1985		European Space Agency probe, Halley's Comet flyby
Suisei	Japan	18 Aug 1985		Halley's Comet flyby

TABLE 11   Winners of U.S. Congressional Space Medal of Honor

ASTRONAUT	DATE OF AWARD	AWARDED BY
Neil Armstrong	1 Oct 1978	President Jimmy Carter
Frank Borman	1 Oct 1978	President Jimmy Carter
Charles Conrad	1 Oct 1978	President Jimmy Carter
John Glenn	1 Oct 1978	President Jimmy Carter
Virgil "Gus" Grissom	1 Oct 1978	President Jimmy Carter (medal awarded posthumously)
Alan Shepard	1 Oct 1978	President Jimmy Carter
John Young	19 May 1981	President Ronald Reagan

TABLE 12 Missions that Failed To Achieve Earth Orbit

MISSION	COUNTRY	LAUNCH
Vanguard TV3	U.S.	6 Dec 1957
Vanguard TV3BU	U.S.	5 Feb 1958
Explorer 2	U.S.	5 Mar 1958
Vanguard TV5	U.S.	28 Apr 1958
Luna 1 precursor	U.S.S.R.	1 May 1958
Vanguard SLV 1	U.S.	27 May 1958
Vanguard SLV 2	U.S.	26 Jun 1958
Explorer 5	U.S.	24 Aug 1958
Luna 1 precursor	U.S.S.R.	24 Sep 1958
Vanguard SLV 3	U.S.	26 Sep 1958
Beacon 1	U.S.	23 Oct 1958
Pioneer 2	U.S.	8 Nov 1958
Luna 1 precursor	U.S.S.R.	26 Nov 1958
Vanguard SLV 5	U.S.	13 Apr 1959
Discoverer 3	U.S.	3 Jun 1959
Vanguard SLV 6	U.S.	22 Jun 1959
Discoverer 4	U.S.	25 Jun 1959
Explorer S1	U.S.	16 Jul 1959
Beacon 2	U.S.	14 Aug 1959
Discoverer 9	U.S.	4 Feb 1960
Discoverer 10	U.S.	19 Feb 1960
Midas 1	U.S.	26 Feb 1960
Explorer S46	U.S.	23 Mar 1960
Echo A1	U.S.	13 May 1960
Discoverer 12	U.S.	29 Jun 1960
Courier 1A	U.S.	18 Aug 1960
Unnamed lunar probe	U.S.	25 Sep 1960
Mars 1 precursor	U.S.S.R.	10 Oct 1960
SAMOS 1	U.S.	11 Oct 1960
Mars 1 precursor	U.S.S.R.	14 Oct 1960
Discoverer 16	U.S.	26 Oct 1960
Solrad 2	U.S.	30 Nov 1960
Transit 3A	U.S.	30 Nov 1960
Explorer S56	U.S.	4 Dec 1960
Explorer S45	U.S.	24 Feb 1961
Kosmos 21 precursor	U.S.S.R.	26 Feb 1961
Discoverer 22	U.S.	30 Mar 1961
Explorer S45A	U.S.	24 May 1961
Discoverer 24	U.S.	8 Jun 1961
Explorer S55	U.S.	30 Jun 1961
Discoverer 27	U.S.	21 Jul 1961
Discoverer 28	U.S.	3 Aug 1961
SAMOS 3	U.S.	9 Sep 1961
Discoverer 33	U.S.	23 Oct 1961
SAMOS 4	U.S.	22 Nov 1961
Discoverer 37	U.S.	13 Jan 1962
Echo	U.S.	15 Jan 1962
Injun 2	U.S.	24 Jan 1962

MISSION	COUNTRY	LAUNCH
LOFTI	U.S.	24 Jan 1962
SECOR 1A	U.S.	24 Jan 1962
Solrad 4A	U.S.	24 Jan 1962
Surcal 1A	U.S.	24 Jan 1962
ERS 1	U.S.	12 Apr 1962
Solrad 4B	U.S.	26 Apr 1962
AC 1	U.S.	8 May 1962
ANNA 1A	U.S.	10 May 1962
P35-1	U.S.	23 May 1962
Mariner 1	U.S.	22 Jul 1962
ERS 3	U.S.	17 Dec 1962
ERS 4	U.S.	17 Dec 1962
Midas 6	U.S.	17 Dec 1962
Luna 4 precursor	U.S.S.R.	3 Feb 1963
Discoverer 59	U.S.	28 Feb 1963
Discoverer 60	U.S.	18 Mar 1963
Hitchhiker 1	U.S.	18 Mar 1963
Transit 5A2	U.S.	5 Apr 1963
Discoverer 62	U.S.	26 Apr 1963
P35-4	U.S.	26 Apr 1963
ERS 7	U.S.	12 Jun 1963
ERS 8	U.S.	12 Jun 1963
MIDAS 8	U.S.	12 Jun 1963
P35-5	U.S.	27 Sep 1963
Discoverer 72	U.S.	9 Nov 1963
Kosmos 27 precursor	U.S.S.R.	4 Mar 1964
Explorer S66	U.S.	19 Mar 1964
Discoverer 76	U.S.	24 Mar 1964
Kosmos 60 precursor	U.S.S.R.	9 Apr 1964
Radose 5E2	U.S.	21 Apr 1964
Transit 5BN3	U.S.	21 Apr 1964
Kosmos 60 precursor	U.S.S.R.	29 Apr 1964
ERSS	U.S.	25 Jun 1964
AC 3	U.S.	30 Jun 1964
Transtage 1	U.S.	1 Sep 1964
I. SAMOS 12	U.S.	8 Oct 1964
OV1 1	U.S.	21 Jan 1965
AC 5	U.S.	2 Mar 1965
OV1 3	U.S.	27 May 1965
I. SAMOS 20	U.S.	12 Jul 1965
OSO-C	U.S.	25 Aug 1965
Starad 2	U.S.	2 Sep 1965
G-6 Target	U.S.	25 Oct 1965
P35-14	U.S.	6 Jan 1966
Discoverer 106	U.S.	3 May 1966
G-9 Target A	U.S.	17 May 1966
OV1 7	U.S.	13 Jul 1966
IDCSP	U.S.	26 Aug 1966
Lambda 4S1	Japan	26 Sep 1966
Lambda 4S2	Japan	20 Dec 1966

TABLE 12   141

MISSION	COUNTRY	LAUNCH
OV3-5	U.S.	31 Jan 1967
Lambda 4S3	Japan	13 Apr 1967
T. SAMOS 5	U.S.	26 Apr 1967
ESRO 2A	ESA[a]	29 May 1967
OV1 11	U.S.	27 Jul 1967
Zond 4 precursor	U.S.S.R.	22 Nov 1967
Zond 5 precursor	U.S.S.R.	22 Apr 1968
Nimbus B	U.S.	18 May 1968
SECOR 10	U.S.	18 May 1968
RM 18	U.S.	16 Aug 1968
SECOR 11	U.S.	16 Aug 1968
SECOR 12	U.S.	16 Aug 1968
UV Radiometer	U.S.	16 Aug 1968
Grid Sphere Drag	U.S.	16 Aug 1968
LCS 3	U.S.	16 Aug 1968
Lidos	U.S.	16 Aug 1968
Orbis Cal 1	U.S.	16 Aug 1968
OV5-8	U.S.	16 Aug 1968
Radcat	U.S.	16 Aug 1968
Intelsat 3F1	International[b]	18 Sep 1968
STV 1	ELDO[c]	30 Nov 1968
Zond 7 precursor	U.S.S.R.	5 Jan 1969
Zond 7 precursor	U.S.S.R.	8 Jan 1969
Kosmos 419 precursor	U.S.S.R.	27 Mar 1969
Luna 15 precursor	U.S.S.R.	15 Apr 1969
Luna 15 precursor	U.S.S.R.	12 Jun 1969
X1	United Kingdom	28 Jun 1969
STV 2	ELDO[c]	2 Jul 1969
Pioneer E	U.S.	27 Aug 1969
TETR-C	U.S.	27 Aug 1969
Lambda 4S4	Japan	22 Sep 1969
SKW 1 precursor	China	1 Nov 1969
Luna 16 precursor	U.S.S.R.	19 Feb 1970
X2	United Kingdom	4 Mar 1970
Orba	United Kingdom	2 Sep 1970
OAO-B	U.S.	30 Nov 1970
Discoverer 139	U.S.	17 Feb 1971
Mariner H	U.S.	8 May 1971
BMEWS 5	U.S.	4 Dec 1971
D2A Polaire	France	5 Dec 1971
T. SAMOS 34	U.S.	16 Feb 1972
T. SAMOS 36	U.S.	20 May 1972
D5A	France	21 May 1973
D5B	France	21 May 1973
T. SAMOS 40	U.S.	26 Jun 1973
ITOS-E	U.S.	16 Jul 1973
SKW 4 precursor	China	18 Sep 1973
Sphinx	U.S.	11 Feb 1974
Viking test	U.S.	11 Feb 1974
SKW 3 precursor	China	12 Jul 1974

MISSION	COUNTRY	LAUNCH
SKW 4 precursor	China	4 Nov 1974
Intelsat 4F6	International[b]	20 Feb 1975
Soyuz 18 precursor	U.S.S.R.	5 Apr 1975
Luna 24 precursor	U.S.S.R.	13 Oct 1975
DAD	U.S.	5 Dec 1975
Corsa	Japan	4 Feb 1976
DMSP	U.S.	18 Feb 1976
Interkosmos 15A	Interkosmos[d]	3 Jun 1976
OTS 1	ESA[a]	13 Sep 1977
Intelsat 4AF5	International[b]	29 Sep 1977
DSCS II-9	U.S.	25 Mar 1978
DSCS II-10	U.S.	25 Mar 1978
SKW 9A	China	30 Jul 1979
Rohini	India	10 Aug 1979
Amsat	W. Germany	23 May 1980
CAT 2	ESA[a]	23 May 1980
Feuerrad	W. Germany	23 May 1980
Navstar 7	U.S.	18 Dec 1981
Marecs B	ESA[a]	10 Sep 1982
Sirio 2	ESA[a]	10 Sep 1982
Soyuz T10 precursor	U.S.S.R.	26 Sep 1983
Eutelsat 3	ESA[a]	12 Sep 1985
Spacenet 3	U.S.	12 Sep 1985
Spartan Halley	U.S.	28 Jan 1986
STS-51L *(Challenger)*	U.S.	28 Jan 1986
TDRS-B	U.S.	28 Jan 1986
Big Bird 24	U.S.	18 Apr 1986
GOES-G	U.S.	3 May 1986
Intelsat 5F14	International[b]	30 May 1986
Rohini 4	India	24 Mar 1987
FLTSATCOM 6	U.S.	26 Mar 1987

[a] **European Space Agency.**

[b] **International venture by the Intelsat Organization.**

[c] **European Launcher Development Organization.**

[d] **International venture by Interkosmos.**

TABLE 13   Space Exploration Fatalities

ASTRONAUT/COSMONAUT	COUNTRY	DATE	NOTES
Roger Chaffee	U.S.	27 Jan 1967	Mission: Apollo 1
Georgi Dobrovolsky	U.S.S.R.	29 Jun 1971	Mission: Soyuz 11
Virgil "Gus" Grissom	U.S.	27 Jan 1967	Mission: Apollo 1
Vladimir Komarov	U.S.S.R.	24 Apr 1967	Mission: Soyuz 1
Christa McAuliffe	U.S.	28 Jan 1986	Mission: STS 51L
Ronald McNair	U.S.	28 Jan 1986	Mission: STS 51L
Ellison Onizuka	U.S.	28 Jan 1986	Mission: STS 51L
Viktor Patsayev	U.S.S.R.	29 Jun 1971	Mission: Soyuz 11
Judith Resnik	U.S.	28 Jan 1986	Mission: STS 51L
Francis Scobee	U.S.	28 Jan 1986	Mission: STS 51L
Michael Smith	U.S.	28 Jan 1986	Mission: STS 51L
Vladislav Volkov	U.S.S.R.	29 Jun 1971	Mission: Soyuz 11
Edward White	U.S.	27 Jan 1967	Mission: Apollo 1

Apollo 1: **Killed in cabin fire during test firing of rocket.**

Soyuz 11: **Killed during re-entry, when a valve accidentally opened
and released the cabin atmosphere.**

STS 51L: **Killed when the space shuttle *Challenger* exploded
73 seconds after lift-off.**

5

Directory of Organizations

THIS CHAPTER IS AN INTERNATIONAL space directory. It lists organizations engaged in various space exploration activities. The organizations are listed alphabetically by name, and each listing includes address, phone, and a description of the organization's activities. For foreign organizations, U.S. offices are listed.

An important acronym used in this chapter is CCDS, or Center for Commercial Development of Space. CCDSs are NASA-sponsored centers where government, higher education, and industry come together to perform focused research that will lead to increased investment and commercialization in space. NASA grants support to each project for five years. The industrial sector matches the support and participates with the CCDSs in joint research leading to commercial space ventures.

Aerospace Industries Association
1250 Eye Street, NW
Washington, DC 20005
(202) 371-8400
Don Fuqua, President

Established in 1919, this nonprofit trade association represents the nation's manufacturers of commercial, military, and business aircraft; helicopters; aircraft engines; missiles; spacecraft; and related components and equipment. The association utilizes the expertise of executives in the aerospace industry to carry common concerns to

Congress, the media, and the public in an effort to establish public awareness of the complex dimensions of the aerospace industry. The association's 50 members represent about 90 percent of U.S. aerospace sales.

PUBLICATIONS: Bimonthly newsletter.

American Astronautical Society
6352 Rolling Mill Place, Suite 102
Springfield, VA 22152
(703) 866-0020
E. Larry Heacock, President

Incorporated in the state of New York in 1954, the society is dedicated to the advancement of the astronautical sciences, space flight engineering, and the astronautical arts. The goals of the society are furthered by international exchange of ideas and information through conferences, meetings, and symposia.

PUBLICATIONS: *Space Times,* bimonthly; *Journal of the Astronautical Sciences,* quarterly; various books through Univelt Publishers, San Diego, California.

American Institute of Aeronautics and Astronautics
370 L'Enfant Promenade, SW
Washington, DC 20024-2518
(202) 646-7400
Cort Durocher, Executive Director

A professional society devoted to science and engineering in aviation and space. The institute's purposes are to advance the arts, sciences, and technology of aeronautics and astronautics and to nurture and promote the professionalism of those engaged in these pursuits. The institute has 58 technical committees that cover all aspects of aerospace. The institute sponsors technical meetings, conferences, and short courses, and it has an active Honors and Awards program that recognizes outstanding achievements in aviation and space. The institute conducts a vigorous public policy program and frequently testifies before congressional committees on aerospace hearings. Membership: 44,000.

PUBLICATIONS: *Aerospace America,* monthly; six technical journals; two book series; and a student journal.

California Space Institute, A-021
Scripps Institution of Oceanography
University of California, San Diego
La Jolla, CA 92093
(619) 534-6381
Sally K. Ride, Director

Founded in 1979, the California Space Institute (CalSpace) is a state-wide research unit that conducts and supports space-related research throughout the University of California system. CalSpace's mission is to identify and develop space sciences and technologies that are beneficial to California's citizens and economy. To that end, CalSpace acts as a liaison between academia, industry, and government. CalSpace also operates a small grant program that allocates funding to researchers in astrophysics, space science, climatology, satellite remote sensing, and space technology. In 1989, NASA awarded CalSpace funding to organize a consortium of University of California campuses—San Diego, Los Angeles, Berkeley—and develop a Space Grant College and Fellowship program. The program, designed to last at least five years, will provide fellowships and create new curricula in space sciences and engineering on the three campuses. Further, CalSpace scientists conduct basic research in fields ranging from satellite-based climate studies to tele-robotic applications. The development of improved instrumentation and space propulsion systems as well as the design of artificial intelligence, automation, and robotic systems for NASA's space station are among CalSpace's research areas.

Center for Advanced Materials
Battelle Columbus Laboratories
505 King Avenue
Columbus, OH 43201-2693
(614) 424-6376
Frank J. Jelinek, Director

A CCDS established in 1985, this center conducts research in advanced materials including polymers, catalysts, electronic materials, metals, ceramics, and superconductors. Commercial applications of these materials are selected for study in microgravity, with the objective of improving the characteristics of the commercial product, improving the process by which the product is made, or gaining additional information about the material that could improve Earth-based processing.

PUBLICATIONS: Scientific papers.

Center for Advanced Space Propulsion
University of Tennessee Space Institute
P.O. Box 1385
Tullahoma, TN 37388
(615) 454-9294
George Garrison, Director

Established in 1987, this CCDS is engaged in the development of innovative, efficient, and highly reliable space propulsion systems. The center's mission is to strengthen U.S. competitiveness in space propulsion technology. Research is focused on three basic areas: economic and reliable access to low Earth orbits, effective means for orbital transfers, and high-performance systems for lunar and interplanetary flights. More specifically, the center's projects focus on five major subjects: advanced chemical propulsion, expert systems, microgravity fluid management, electric propulsion, and rocket engine materials processing. The center's partners so far include Boeing Aerospace Company, Rockwell's Rocketdyne Division, Sundstrand Corporation, and Saturn Corporation.

PUBLICATIONS: Scientific papers.

**Center for Autonomous
and Man-Controlled Robotic and Sensing Systems**
P.O. Box 8618
Ann Arbor, MI 48107
(313) 994-1200
Charles J. Jacobus, Director

This CCDS was established in 1987 with a $5 million grant from NASA. The center focuses its research in three main areas: automated construction, repair, and maintenance of orbital and remote planetary systems; automated biological processing and biological systems for long-duration life support; and automated robotics for ground-based servicing, construction, and mining systems. Current projects include in-orbit serviceable satellites, advanced space robot design, mobile robotics systems, automated microgravity experimentation, and automated screening systems and plant cell culture.

PUBLICATIONS: Scientific papers.

Center for Bioserve Space Technologies
University of Colorado at Boulder
Campus Box 429
Boulder, CO 80309
(303) 492-1005
Marvin Luttges, Director

A CCDS established in 1987. The center's projects can be grouped into four categories:

1. Bioproducts/bioprocessing focuses on forming and manipulating biomaterials in the microgravity of space. Results from this research may be used to produce artificial skin, tendons, blood vessels, and corneas.
2. Studies of biomedical isomorphisms, which are the numerous degenerative changes that occur in astronauts exposed to the reduced gravity and radiation of space. Results from research such as on the bone material loss that affects

astronauts in space is being used to model osteoporosis processes that occur on Earth.

3. Controlled ecological life-support systems research involves the investigation of biologically based technologies that will permit efficient, effective, and simple waste management, water reclamation, and production of food and oxygen in space.

4. Hardware development is done to produce tools that support bioserve projects.

PUBLICATIONS: Scientific papers.

Center for Cell Research
South Frear Building
Pennsylvania State University
University Park, PA 16802
(814) 865-2407
Sylvia Stein, Executive Director

A CCDS established in 1987. Its main goals are definition of the fundamental mechanisms of mammalian cell function on Earth and in space and commercialization of the findings in cooperation with business and industry. The center's three major projects focus on the musculoskeletal system (bone and hematopoiesis), the light-endocrine-immune axis, and bioprocessing. This research will lead to a definition of microgravity-caused modifications in cells and tissues and to the establishment of the relationship between these space-caused modifications and diseases of humans on Earth. The research will also then shed light on commercial concerns such as aging, anemia, immune functions, and muscle atrophy.

PUBLICATIONS: Scientific papers.

Center for Commercial Crystal Growth in Space
Clarkson University
Potsdam, NY 13676
(315) 268-6446
Bill Wilcox, Director

A CCDS established in 1986. The center's primary goal is the development of commercial crystal growth in space, with emphasis on the growth of single crystals of higher perfection than can be grown on Earth. Applications of this venture include electronics, optical systems, infrared detectors, and chemical separations. Research is performed by three teams, each concentrating in one of these areas: melt growth, vapor growth, and solution growth. The teams are performing ground-based experiments and planning flight experiments.

PUBLICATIONS: Scientific papers.

Center for Commercial Development of Space Power
Space Power Institute
Auburn University
Auburn, AL 36849-5320
(205) 844-5894
Raymond Askew, Director

A CCDS established in 1987. The center's objectives are to identify the critical technological impediments to the economic deployment of power systems in space; to advance these technologies when they become available; and to develop new products to meet the power generation, storage, conditioning, and distribution needs of commercial space users. The center's research areas are power distribution, control, and management; discrete energy and power sources; advanced power system sensors; power conditioning; and power transmission. The center's research will provide the basis for developing complete power system packages tailored to specific user needs.

PUBLICATIONS: Scientific papers.

Center for Excellence in Space Data and Information Sciences
Code 630.5, Goddard Space Flight Center
Greenbelt, MD 20771
(301) 286-4403
Raymond E. Miller, Director

Established in 1988, this new research center in advanced computer sciences was developed jointly by NASA, the Universities Space Research Association, and the University of Maryland to support NASA's Earth and space science programs. The center will bring together computer scientists from leading university, industrial, and government laboratories to focus attention on accessing, processing, and analyzing data from space observing systems; to collaborate with NASA's space and Earth scientists; and to conduct other computer science research that has applications in space and Earth sciences.

Center for Macromolecular Crystallography
University of Alabama at Birmingham
THT-Box 79, University Station
Birmingham, AL 35294
(205) 934-5329
Charles E. Bugg, Director

A CCDS established in 1985, the center focuses on biotechnology research in the microgravity of space. Research is focused on developing crystallization experiments and designing the hardware, techniques, and facilities for growing protein crystals in space. The center also offers other protein crystallography services to the pharmaceutical, chemical, and biotechnology industries. The center has performed protein crystal growth experiments on six past space shuttle missions. A series of further experiments is planned for shuttle flights through the mid-1990s.

PUBLICATIONS: Scientific papers.

Center for Mapping
Ohio State University
1958 Neil Avenue
Columbus, OH 43210-1247
(614) 292-6642
John Bossler, Director

A CCDS established in 1986. The research at this center is driven by commercial needs in land, water, and farm management; energy and power production; digital mapping; information systems; and disaster assessment. Research efforts have included forecasting storm surge levels, analyzing the impact of drought conditions in the Midwest, anticipating satellite orbits, monitoring gas leaks, predicting the effects of erosion, and tracking ocean currents.

PUBLICATIONS: Scientific papers.

Center for Materials for Space Structures
Case Western Reserve University
White Building, Room 418
Cleveland, OH 44106
(216) 368-4222
John F. Wallace, Director

A CCDS established in 1987. The center's goal is to provide materials for space structures that can be made and assembled in space and can withstand the space environment. The center focuses on five specific project areas: films and foamed materials, organic and inorganic composites, and organic and inorganic coatings. The center is striving to develop space structural materials with a life expectancy of at least 30 years and resistance to environmental and mechanical attack in low Earth orbit. NASA funding of the center is set aside for five years at the rate of $800,000 per year. Case Western Reserve University is also supporting the center for its first four years, in addition to support provided by the state of Ohio.

PUBLICATIONS: Scientific papers.

Center for Space and Geosciences Policy
University of Colorado
Campus Box 361
Boulder, CO 80309-0361
(303) 492-1171
Radford Byerly, Director

The center provides an interdisciplinary policy focus that complements two of the university's strongest technical areas: space science and technology, and Earth systems science. The center's space policy research is aimed at addressing the need for independent space policy research and analysis and providing a university-based source for studies of the legal, political, commercial, and international issues arising from humankind's exploration and use of outer space. In the geosciences area, the center's policy research addresses such topics as the "greenhouse effect" and its implications for existing agricultural, hydrological, and economic infrastructures.

PUBLICATIONS: Research reports and papers.

Center for Space Automation and Robotics
University of Wisconsin
1357 University Avenue
Madison, WI 53715
(608) 262-5524
Neil Duffie, Director, Astrobotics
Theodore Tibbitts, Director, Astroculture
Gerald Kulcinski, Director, Astrofuel

A CCDS established in 1986 with a $5 million grant from NASA, the center is dedicated to developing technology for the commercialization of automated systems for use in space and on Earth. Research is focused in three major areas: astrobotics, astroculture, and astrofuel. Astrobotics deals with creating robotic technologies capable of enhancing humankind's ability to live, travel, and explore in space. Astroculture deals with developing automated plant growth facilities for space. Astrofuel focuses on the mining, processing, and transporting of helium-3, a valuable source of safe, clean, and reliable fusion fuel. (The Moon, for example, has enough helium-3 on its surface to meet Earth's energy needs for centuries if it were transported to Earth.)

PUBLICATIONS: Scientific papers.

Center for Space Power
Texas A & M University System
College Station, TX 77843-3118
(409) 845-8768
A. D. Patton, Director

A CCDS established in 1987, the center identifies and conducts research to enhance space power systems technologies. Research interests include power systems for individually designed applications, such as a communication satellite or the space shuttle, and for future needs, such as space stations, human-tended free fliers, lunar bases, and other powered installations. Examples of short-term projects include improving the individual components currently used in space power systems, such as batteries, fuel cells, solar panels, and power conditioning. Long-term projects address the development of new power production systems using photovoltaics, solar dynamics, and nuclear systems—or a combination of these—including power distribution through hardwiring or by microwave or laser transmission.

PUBLICATIONS: Scientific papers.

Center for Space Processing
P.O. Box 6309, Station B
Nashville, TN 37235
(615) 322-7047
Tony Overfelt, Director

A CCDS established in 1985, the center is located on the Vanderbilt University campus. Major research involves microgravity experimentation in container-less processing, directional solidification, casting, and cold welding. Studies are being conducted on four materials: metals, alloys, ceramics, and glass. The results of these efforts will be seen in the application of research to Earth-based manufacturing processes and will provide the basis for future commercial exploration of the space environment and resources.

PUBLICATIONS: Scientific papers.

Center for Space Vacuum Epitaxy
University of Houston
Houston, TX 77204-5507
(713) 747-3701
Alex Ignatiev, Director

A CCDS established in 1986. The center's primary research focuses on exploring epitaxial thin-film growth and materials purification in the ultra-vacuum of space. An ultra-high vacuum allows for the controlled deposition of unique thin films by epitaxial growth. Researchers consider vacuum epitaxy to be the most powerful technique available for synthesizing new thin-film electronic, superconducting, and magnetic materials and devices. Unique epitaxial procedures developed at this center can be used to manufacture such products, which possess various enhanced qualities. The center's future efforts will concentrate on adapting molecular beam epitaxy and chemical beam epitaxy technologies to the space ultra-vacuum environment and on the development of semiconductor thin-film materials and devices.

PUBLICATIONS: Newsletter to members and staff.

Consortium for Materials Development in Space
University of Alabama, Research Institute Building
Huntsville, AL 35899
(205) 895-6620
Charles Lundquist, Director

A CCDS established in 1985, the consortium focuses on investigations in space as a means to develop new materials and processes. This approach covers commercial materials development that benefits from the unique attributes of space, commercial applications of the physical chemistry that occurs at the surface of a new material and how materials are transported to it, and prompt and frequent experiments and operations in space. The consortium's most significant accomplishment so far has been a breakthrough in high-temperature superconductors. The consortium developed the first materials that conduct electricity, without resistance, at temperatures "hotter" than liquid nitrogen, -196°C (-321°F).

PUBLICATIONS: Scientific papers.

Cosmic
University of Georgia
382 Broad Street
Athens, GA 30602
(404) 542-3265
John Gibson, Director

Established in 1966 to facilitate the transfer of technology from NASA to the general public via the medium of computer software, Cosmic is the distribution center for computer software created under NASA funding. The Cosmic inventory contains approximately 1,200 different computer programs in fields such as aerodynamics, artificial intelligence, computer-assisted design (CAD) and computer-assisted manufacturing (CAM), composites, control systems, finite element analysis, heat transfer and fluid flow, image processing, optics, project management, reliability, satellite communications, scientific visualization, and turbine engineering. Cosmic's services include a custom search of the inventory to identify programs that might help solve a particular problem. Cosmic is sponsored by NASA and is affiliated with NASA's Technology Utilization Program.

PUBLICATIONS: Annual catalog, quarterly newsletter, and promotional literature.

Department of Transportation
Office of Commercial Space Transportation
400 Seventh Street, SW (S-50)
Washington, DC 20590
(202) 366-5770
Stephanie Lee-Miller, Director

Established in 1984 by the Commercial Space Launch Act of 1984, the office is responsible for the promotion and regulation of commercial launch activities. Its activities include promoting, encouraging, and facilitating commercial space transportation, and licensing and regulating all U.S. commercial launch activities to ensure that they are conducted safely and responsibly.

PUBLICATIONS: Annual report.

European Space Agency (ESA)
955 L'Enfant Plaza, SW, Suite 7800
Washington, DC 20024
(202) 488-4158
Reimar Lüst, Director General

ESA was established in 1975 by the founding members: Belgium, Denmark, Federal Republic of Germany, France, Ireland, Italy, the Netherlands, Spain, Sweden, Switzerland, and the United Kingdom. Austria, Norway, and Finland joined in 1987. ESA's purpose is to provide for and promote, for exclusively peaceful purposes, cooperation among European states in space research and technology and their space applications with a view to their being used for scientific purposes and for space application systems. One of ESA's main tasks is to prepare for its member nations, at regular intervals, long-term plans proposing the direction European space research should take. ESA also plays a coordination role by following closely the national programs of its member states and, whenever feasible, integrating these into its own programs. ESA is essentially an oversight organization and does not develop or manufacture its own spacecraft. Development work is carried out by the industries of the member states under the watchful eye of ESA staff.

PUBLICATIONS: ESA *Bulletin, Columbus Logbook,* ESA *Journal, Earth Observation Quarterly,* and *Reach for the Skies,* all quarterly.

Goddard Institute for Space Studies
2880 Broadway
New York, NY 10025
(212) 678-5500
James Hansen, Director

Established in 1961, the institute is a New York division of Goddard Space Flight Center's Space and Earth Sciences Directorate. Its purpose is to do basic research in space and earth sciences in support of NASA Goddard Space Center programs. Current research at the institute is aimed at a broad study of Global Change. Global Change is an interdisciplinary research initiative addressing natural and man-made changes in our environment that occur on time scales of decades and affect the habitability of our planet. Program areas at the institute include global climate, biogeochemical cycles, planetary atmospheres, Earth observations, and related interdisciplinary studies.

International Space University
955 Massachusetts Avenue
7th Floor
Cambridge, MA 02139
(617) 354-1987
Todd B. Hawley, Chief Executive Officer

Established in 1987. A nonprofit international graduate education program for space research and development. The first educational institution of its kind, the university was founded to provide graduate-level young professionals, who demonstrate academic excellence and leadership qualities, with summer session programs in space education. Each year, the summer program is highlighted by the completion of a design project with the aim of offering students a chance to integrate course work into real-life applications. The results of these projects are returned to the respective countries of students for further development and implementation at a national level. The university is in the process of adding satellite campuses around the world.

PUBLICATIONS: *The Universe*, quarterly newsletter.

NASA Headquarters
400 Maryland Avenue
Washington, DC 20546
(202) 453-1030
Admiral Richard Truly, Administrator

This office exercises management over all space flight centers, research centers, and other installations that constitute the National Aeronautics and Space Administration. Headquarters responsibilities cover the determination of programs and projects; the establishment of management policies, procedures, and performance criteria; the evaluation of progress; and the review and analysis of all phases of the aerospace program.

NASA
Ames Research Center
Moffett Field, CA 94035
(415) 604-5000
Dale Compton, Director

Founded in 1940, Ames Research Center specializes in scientific research, exploration, and applications with the aim of creating new technology. Major programs are in computer science and applications, computational and experimental aerodynamics, flight simulation, flight research, hypersonic aircraft, rotorcraft and powered-lift technology, aeronautical and space human factors, life sciences, space sciences, Solar System exploration, airborne science and applications, and infrared astronomy. The center supports the space shuttle and various military and civil aviation programs.

NASA
Central Operation of Resources for Educators
Lorain County JVS
15181 Route 58 South
Oberlin, OH 44074
(216) 774-1051
Tina Salyer, Coordinator

A national distribution center for aerospace education materials. Materials available include videotape, slide, and filmstrip programs that chronicle NASA's state-of-the-art research and technology. Subject areas include life sciences, physical science, astronomy, energy, Earth resources, environment, mathematics, and career education. Also available are single-subject programs covering all areas of NASA's operations. Through the use of these curriculum supplement materials, teachers can provide their students with the latest in aerospace information.

PUBLICATIONS: Annual catalog.

NASA
Dryden Flight Research Facility
P.O. Box 273
Edwards Air Force Base, CA 93523
(805) 258-3311
Martin Knutson, Site Manager

A flight research arm of NASA and the military. Facilities include a laboratory for testing complete aircraft and structural components and flight instrumentation, a flight systems laboratory for avionics systems, flow visualization laboratories, a data analysis facility for processing flight research data, and a facility for testing remotely piloted vehicles. The Dryden facility was actively involved in the approach and landing tests of the space shuttle *Enterprise* and continues to support shuttle landings from space.

NASA
Goddard Space Flight Center
Greenbelt, MD 20771
(301) 286-6255
John Klineberg, Director

The mission of this flight center is to extend the horizons of human knowledge not only about the Solar System and the universe but also about Earth and its environment. The mission is being accomplished through scientific research centered in six space and Earth science laboratories and in the management, development, and operation of several near-Earth space systems such as the Hubble Space Telescope, the Cosmic Background Explorer, the Gamma Ray Observatory, and the Upper Atmosphere Research Satellite. Goddard is also the nerve center of NASA's worldwide ground and space-borne communications network. In addition, Goddard is involved with the development of the space station. Goddard's charter is to develop the detailed design, construction, testing, and evaluation of the automated free-flying polar platform and provisions for instruments and payloads to be attached externally to the space station.

NASA
Jet Propulsion Laboratory (JPL)
4800 Oak Grove Drive
Pasadena, CA 91109
(818) 354-4321
Lew Allen, Director

JPL is a government-owned facility operated by the California Institute of Technology under contract. JPL is engaged in activities associated with deep space automated scientific missions. Examples of the activities are engineering subsystem and instrument development and studies of data reduction and analysis required by deep space flight. Current NASA flight projects under JPL management include Voyager, Galileo, Magellan, and the Mars Observer. JPL is also involved in developing a wide-field and planetary camera for the Hubble Space Telescope and an imaging radar for the space shuttle.

NASA
Johnson Space Center (JSC)
Houston, TX 77058
(713) 483-6511
Aaron Cohen, Director

Established in 1961 as NASA's primary center for design, development, and testing of spacecraft and associated systems for manned space flight; selection and training of astronauts; planning and execution of manned missions; and extensive participation in the medical, engineering, and scientific experiments carried aboard space flights. In addition to program management responsibility for the space shuttle program, JSC has major responsibility for the development of the space station.

NASA
Kennedy Space Center (KSC)
Kennedy Space Center, FL 32899
(407) 867-2468
Lieutenant General (Retd.) Forrest S. McCartney, Director

Established in the 1960s to serve as a launch site for the Apollo lunar missions. KSC is NASA's primary center for the testing, maintenance, and launch of manned and unmanned spacecraft. KSC is responsible for the assembly, maintenance, and launch of space shuttles and their payloads, their landing operations, and their preparation between missions.

NASA
Langley Research Center
Hampton, VA 23665-5225
(804) 864-6123
Richard H. Petersen, Director

The primary mission of Langley Research Center is basic research in aeronautics and space technology. Major research fields include aerodynamics, materials, structures, controls, information systems, acoustics, aeroelasticity, and atmospheric sciences. About 60 percent of research is aeronautical, aimed at improving today's aircraft and developing concepts and technology for the aircraft of the future. Langley seeks to develop aircraft that fly faster, farther, and safer and are more maneuverable, quieter, cheaper, and more energy efficient than today's aircraft. The remaining 40 percent of Langley's attention is focused on space programs. Langley develops technologies for advanced space transportation systems and laser techniques for space applications and for large space systems such as those associated with the space station *Freedom.*

NASA
Lewis Research Center
21000 Brookpark Road
Cleveland, OH 44135
(216) 433-2899
Lawrence Ross, Director

Established in 1941, the center has responsibility for developing the largest space power system ever designed, to provide the electrical power necessary for life-support systems and research experiments aboard the space station *Freedom.* In addition, the center will support the space station in other major areas such as auxiliary propulsion systems and communications.

NASA
Marshall Space Flight Center
Huntsville, AL 35812
(205) 544-2121
T. J. Lee, Director

Established in 1960, the center provides the space shuttle's engines, external tank, and solid rocket boosters. The center also plays a key role in the development of payloads for the shuttle. For example, the center helped develop *Spacelab*, a multiuse workshop already in orbit. Three successful missions have already used *Spacelab* under the center's direction. The center will also be in charge of future missions to *Spacelab*. Marshall also has management responsibility for the Hubble Space Telescope; the Tethered Satellite System, to be placed in orbit by 1991; the Orbital Maneuvering Vehicle, which will be an unmanned vehicle for performing numerous activities such as moving satellites from one orbit to another; and the U.S. modules of the space station *Freedom*. The U.S. modules include living quarters, the U.S. laboratory, logistics elements, node structures that connect modules, fluids and environmental control, life-support subsystems, internal thermal control, and audio-video systems. Finally, Marshall is investigating an unmanned, cargo version of the space shuttle.

PUBLICATIONS: *Research and Technology,* annual report; *Marshall Star,* internal weekly newsletter.

NASA
Michoud Assembly Facility
P.O. Box 29300
New Orleans, LA 70189
(504) 257-2601
John W. Hill, Manager

The mission of the Michoud facility includes systems engineering, engineering design, fabrication, assembly, and other work related to the space shuttle's external tank.

NASA
Office of Aeronautics, Exploration, and Technology
400 Maryland Avenue
Washington, DC 20546
(202) 453-2693
Arnold D. Aldrich, Associate Administrator

This office was established in 1963 as the Office of Advanced Research and Technology. The name was changed to the Office of Aeronautics and Space Technology in 1972; the Office of Exploration was created in 1987; and Aeronautics, Exploration, and Technology were merged in 1990. The office is responsible for the planning, direction, execution, evaluation, documentation, and dissemination of the results of NASA's research and technology development programs. The programs are conducted primarily to demonstrate the feasibility of any concept, structure, or component that may have general application to the nation's aeronautical and space objectives.

PUBLICATIONS: Research and technology report, released periodically; annual report.

NASA
Office of Commercial Programs
600 Independence Avenue
Washington, DC 20546
(202) 557-9022
James T. Rose, Assistant Administrator

This office provides the focus within NASA for supporting the expansion of U.S. private sector investment and involvement in civil space activities. The office is responsible for programs supporting new, high-technology commercial space ventures; for the commercial application of existing aeronautics and space technology; and for expanding commercial access to available NASA capabilities and services. The office manages the Centers for Commercial Development of Space.

NASA
Office of Space Flight
Code M, 600 Independence Avenue
Washington, DC 20546
(202) 453-2015
William B. Lenoir, Associate Administrator

This office is responsible for developing and applying a capability that will permit humans to explore space and perform missions leading to increased knowledge and improved quality of life on Earth. To achieve these goals, the office directs the development of space transportation and the required support systems for humans to execute missions in space. The office is also responsible for scheduling space shuttle flights, developing financial plans and pricing structures, providing services to users, managing expendable launch vehicle services and upper stages, and managing NASA's advanced program activities.

NASA
Office of Space Operations
Code T, 600 Independence Avenue
Washington, DC 20546
(202) 453-2019
Charles T. Force, Associate Administrator

This office is responsible for all activities incident to the tracking of aeronautical research aircraft, space launch vehicles, and spacecraft and for the acquisition and distribution of technical and scientific data from such craft. The office is also responsible for managing NASA's communications systems and for operational data systems and services. The office develops plans for management to meet the requirements of the new era of space operations.

NASA
Office of Space Science and Applications
Code E, 600 Independence Avenue
Washington, DC 20546
(202) 453-1409
Leonard A. Fisk, Associate Administrator

This office is responsible for NASA's automated space flight program directed toward scientific investigations of the Solar System using ground-based, airborne, and space techniques including sounding rockets, Earth satellites, and deep-space probes. The office is also responsible for scientific experiments to be conducted by humans in space, for directing the NASA scientific portion of the Spacelab program, and for NASA contact with the Space Science Board of the National Academy of Sciences and other advisory groups.

NASA
Office of Space Station
Code S, 600 Independence Avenue
Washington, DC 20546
(202) 453-2015
William B. Lenoir, Associate Administrator

This office is now a part of the Office of Space Flight. The Office of Space Station is responsible for managing and directing all aspects of the space station program and for achieving the goals established by President Ronald Reagan in his State of the Union address of January 25, 1984. These goals include developing a permanently manned space station by the mid-1990s, encouraging other nations to participate in the space station program, and promoting private-sector investment in space exploration through enhanced space-based operational capabilities.

PUBLICATIONS: *Station Break*, monthly.

NASA
Space Telescope Science Institute
Johns Hopkins Homewood Campus
Baltimore, MD 21218
(301) 338-4757
Eric Chaisson, Head of Education and Public Affairs

The institute plans and conducts science operations for the Hubble Space Telescope. The facility houses offices for scientists and administrative personnel as well as the computer and imaging systems required to evaluate, prepare, and schedule observations for the telescope and to receive, analyze, display, and archive data from the telescope.

NASA
Stennis Space Center
Stennis Space Center, MS 39529
(601) 688-3341
Roy Estess, Director

Stennis's primary mission is testing of the space shuttle's main engine and main orbiter propulsion system, which it has been doing since 1975. Stennis has also evolved into a center of excellence in remote sensing. The center also conducts data systems and commercial utilization studies in support of the space station.

NASA
Wallops Flight Facility
Wallops Island, VA 23337
(804) 824-1579
Keith Coehler, Public Affairs Officer

This facility falls under the administration of the Goddard Space Flight Center. Wallops manages and implements NASA's sounding rocket projects, which use suborbital rocket vehicles to accommodate about 50 scientific missions every year. Launches are conducted at Wallops and other launch facilities throughout the world. Wallops also manages and coordinates NASA's scientific balloon projects using thin-film, helium-filled balloons for about 45 scientific missions each year.

NASA
White Sands Test Facility
P.O. Drawer MM
Las Cruces, NM 88004
(505) 524-5011
Rob R. Tillett, Manager

A research arm of the Johnson Space Center, established in 1964. This facility is where the rocket motors that took humans to the Moon were tested, where the motors of the Viking missions to Mars were tested, and where the advanced propulsion systems of tomorrow are being tested. The facility also tests materials used in spacecraft. Test activities also include component and assembly qualification, flight and test anomaly investigation, research and development on metals ignition, and investigating compatibility of materials and fluids. Finally, space shuttle pilots practice landing on a strip at the facility, which is designated as an alternate landing site for the shuttle.

National Space Development Agency of Japan
Watergate, Suite 570
600 New Hampshire Avenue, NW
Washington, DC 20037
(202) 333-6444
Masanori Nagatomo, Director

The Japanese equivalent of NASA. The agency's space development program is classified into the following categories: (a) the N-1 program uses the N-1 launch vehicle; (b) the GMS-CS-BS program is a meteorological, communications, and broadcasting satellite program that uses U.S. Delta rocket motors; (c) the N-2 program uses the N-2 launch vehicle; (d) the H-1 program uses the H-1 launch vehicle; (e) the H-2 program uses the H-2 launch vehicle; (f) the Space Station Integrated Project is a development program that focuses on the utilization of space for material processing experiments using the space shuttle, *Spacelab,* and, in the future, NASA's space station.

PUBLICATIONS: Monthly report.

National Space Society
International Space Center
922 Pennsylvania Avenue, SE
Washington, DC 20003
(202) 543-3991
Charles Walker, President

A society of space exploration supporters formed in 1987 by merging two pro-space organizations: the National Space Institute, founded in 1974, and the L-5 Society, founded in 1975. Society goals are to promote space exploration, research, development, and habitation; provide broad, visible public support for space exploration; encourage study of mathematics and sciences by creating new space employment opportunities; seek maximum support for space technology research; provide public forums to exchange information; inform the public on space and space technologies; and recognize and honor significant contributions to the space program. Membership: 23,500.

PUBLICATIONS: *Ad Astra,* monthly magazine.

North Texas Space Business Roundtable
10135 Ferndale Road
Dallas, TX 75238
(214) 348-5074
Lenox Carruth, Director

Established 1985. The roundtable's objectives are to support graduate-level research related to space commercialization and to inform the professional, educational, and business communities of the advances and opportunities of space commercialization. To promote these objectives, the roundtable sponsors a monthly luncheon for its members and guests to meet and hear noted speakers.

Space Remote Sensing Center
Building 1103, Suite 118
Stennis Space Center, MS 39529
(601) 688-2509
George May, Director

A CCDS, the center is finding ways to improve the operational productivity and efficiency of those involved in land-use planning and resource management. The center is developing advanced techniques for agricultural crop monitoring, forest mensuration, environmental assessment, facilities monitoring, and land-use planning. Economic benefits are gained directly from improved operational planning, which can lead to lower project cost. In agriculture, for example, the center is developing image processing techniques and computer models to assess vigor and yield for several crops. One product is a crop variability map derived from remotely sensed data. Farmers can use the map to reduce fertilizer costs by fertilizing each soil unit according to the soil's potential to grow a crop.

PUBLICATIONS: Scientific papers and reports.

Spaceweek National HQ
P.O. Box 58172
Houston, TX 77258
(713) 333-3627
Bennet James, Chairman

An organization that coordinates activities to celebrate the nine days from July 16–24 as Space Week. These dates encompass the *Apollo 11* mission to the Moon in 1969. (Neil Armstrong, commander of the mission, became the first human to set foot on the Moon on July 20, 1969.) Spaceweek is a coordinated effort of the U.S. pro-space movement, designed to educate the public about the benefits of space exploration and development while demonstrating public support for such programs.

U.S. House of Representatives
Space and Technology Committee
2321 Rayburn House Office Building
Washington, DC 20515
(202) 225-6371
Robert A. Roe, Chairman

A congressional oversight body of the U.S. space exploration effort.

U.S. House of Representatives
Space Science and Applications Subcommittee
2324 Rayburn House Office Building
Washington, DC 20515
(202) 225-7858
Bill Nelson, Chairman

A congressional oversight body of the U.S. space exploration effort.

United States Space Education Association
746 Turnpike Road
Elizabethtown, PA 17022-1161
(717) 367-3265

Nonprofit organization established in 1973. The association's objectives: to promote the peaceful exploration of outer space through a better educated public; to develop a proposal for the United States to open up the new worlds in space for all humankind in a fair and equitable manner; to stimulate public awareness of the benefits to humankind that accrue from a viable and expanded manned and unmanned space program; and to encourage personal research and understanding of space technology.
PUBLICATIONS: *Space Age Times,* bimonthly.

United States Space Foundation
P.O. Box 1838
Colorado Springs, CO 80901
(719) 550-1000
Richard P. MacLeod, President

Established in 1983. The foundation's goals include enhancing the education of our young people to prepare them for success in a space-, computer-, and technology-oriented society; providing the means for interaction between space professionals—civil, military, commercial, and international—and the general public, so that knowledge about space technology benefits to business and society can be more widely shared; and developing the foundation as an international space resource center that encompasses research, information, computer telecommunications, and space-theme attractions.
PUBLICATIONS: *Spacewatch,* monthly magazine.

World Space Foundation
P.O. Box Y
South Pasadena, CA 91030-1000
(818) 357-2878
Robert L. Staehle, President

Established in 1979. A nonprofit organization that sponsors privately financed space research. The foundation is currently sponsoring two major activities: the Solar Sail Project, which is building a test spacecraft to develop a new propulsion technology that will harness solar winds, and the Asteroid Project, which will augment NASA's efforts to discover and study small bodies in the Solar System. Other foundation activities include defining lunar base options, designing effective and least expensive ways to send the first manned expedition to Mars, and designing a space transport vehicle that utilizes existing technology but reduces the cost of ferrying people into Earth orbit and back.

PUBLICATIONS: *Foundation News,* quarterly; *Foundation Astronautics Notebook,* quarterly; *Under the Stars,* quarterly.

6

Books and Periodicals

THIS CHAPTER PRESENTS AN ANNOTATED LIST of reference books, monographs and government documents, and periodicals pertinent to space exploration.

Reference Books

Hooper, Gordon R. **The Soviet Cosmonaut Team: A Comprehensive Guide to Men and Women of the Soviet Manned Space Programme.** San Diego: Univelt, 1986. 330p. ISBN 0-9511312-0-0.

A reference volume comprising two sections: (1) manned space flight log with time in space; intercosmos countries in order of flight; cosmonaut call signs; cosmonaut extravehicular activities; crewing, training, and selection procedures; and data on training centers; and (2) biographies of about 100 Soviet cosmonauts including those from Eastern Bloc countries and other non-Russian cosmonauts who flew by invitation.

Johnson, Nicholas L. **Handbook of Soviet Lunar and Planetary Exploration** and **Handbook of Soviet Manned Space Flight.** 2d eds. San Diego: Univelt, 1988. 446p. total. ISBN 0-87703-106-1; 0-87703-115-0.

This set of two reference books covers *Vostok, Voskhod, Soyuz, Salyut* space station, and *Soyuz-Salyut* missions. Both volumes include comprehensive appendices of Soviet launch vehicles, facilities, and other tables of information.

King-Hele, D. G. **The RAE Table of Earth Satellites 1957–1986.** 3d ed. New York: Stockton, 1987. 936p. ISBN 0-935859-05-5.

D. G. King-Hele was elected Fellow of the Royal Society of London for this classic tabulation of data on some 17,000 satellites launched until 1986 by the various countries of the space club. The tables contain names of the satellites, the dates they were launched, their orbits, and their purpose. An excellent compilation of data on satellites.

Lewis, Cathleen S., and Dominick A. Pisano, eds. **Air and Space History: An Annotated Bibliography.** New York: Garland, 1988. 571p. ISBN 0-8240-8543-4.

A reference book of works in aeronautics and space sciences with quite thorough annotations. Through the bibliography can be traced the history of the development of these sciences. The book is divided into three major sections: general, air, and space. The general section includes the sources used for compiling the book and works related to the political, economic, cultural, and social issues of air and space. The air section covers everything that flies in Earth's atmosphere, from balloons to helicopters to air ships to aircraft. Finally, the space section spans across the development of rockets to their growth into missiles and then space vehicles. The National Air and Space Museum in Washington, D.C., helped the authors compile the book.

Newkirk, Roland W., Ivan D. Ertel, and Courtney G. Brooks. **Skylab: A Chronology.** Washington: U.S. GPO, 1977. 480p. Stock #3300-00677-2.

Skylab was the first and is so far the only U.S. space station. From its launch on May 14, 1973, to its re-entry and burn-up on July 11, 1979, *Skylab* was host to three different three-man crews who gathered an invaluable wealth of information on the Earth, the Sun, the Solar System, and the stars, not to mention the experience of living for extended periods in the microgravity of space. This book is a chronological history of the *Skylab* space station, starting from 1923, when Hermann Oberth first proposed a manned U.S. space station, to its acceptance scientifically and politically, and then to its development, testing, and launch. A highly comprehensive chronicle of the *Skylab* space station.

Turnill, Reginald, ed. **Jane's Spaceflight Directory 1986.** 2d ed. New York: Jane's Publishing Co., 1986. 453p. ISBN 0-7106-0367-3.

Yet another classic reference work on space exploration. The directory contains vast amounts of historical data on space programs, both manned and unmanned, including military space programs of the United States as well as those of other countries. The book is in its second year of issue. The above issue covers, in addenda, the *Challenger* disaster and its implications for the future of U.S. and international space exploration.

Monographs and Government Documents

Baker, David. **Conquest: A History of Space Achievements from Science Fiction to the Shuttle.** New York: Merrimack Publishers' Circle, 1985. 191p. ISBN 0-947703-00-4.

A chronicle of space flights of the last 20 years, but with emphasis on the post-Apollo era (only about a fourth of the book is devoted to the pre-Apollo era). The book is extremely well illustrated. It also contains valuable data in tabular form, such as information on manned and unmanned flights and on astronauts and cosmonauts and their length of time in space. The text is easy to understand. Critics recommend the book highly for high school students and laypeople in general.

————. **The Rocket: The History and Development of Rocket and Missile Technology.** New York: Crown, 1978. 277p. ISBN 0-517-534404-5.

Traces the development of rocketry, both political and technological. The book includes very absorbing illustrations. Although it contains a description of the space program that is not entirely current, the work can be used by the reader to get an idea of what the space program was a little over a decade ago and how the direction in which it was then headed compares to where it is today.

Belew, Leland F. *Skylab:* **Our First Space Station.** Washington: U.S. GPO, 1977. 176p. Stock #3300-00670-5.

The *Skylab* missions amassed a wealth of information on the Earth, the Sun, and the Solar System. This book tells how that information was collected. The book introduces, aided by excellent illustrations, the *Skylab* missions, their scientific endeavors, their crews, their problems, and their achievements.

Belew, Leland F., and Ernst Stuhlinger. *Skylab:* **A Guidebook.** Washington: U.S. GPO, 1977. 256p. Stock #3300-00508-3.

This illustrated book on the *Skylab* missions sheds light not only on the missions once they had achieved orbit but also on the preparations for the missions. Included are descriptions of the Skylab program, the training of the crew, and preparations of the scientific and other on-board equipment.

Benson, Charles D., and William Barnaby Faherty. **Moonport: A History of Apollo Launch Facilities and Operations.** Washington: U.S. GPO, 1978. 633p. Stock #3300-00740-0.

The purpose of the Apollo program was to take man to the Moon. With such an exciting purpose, other aspects of the program were often overshadowed. In *Moonport: A History of Apollo Launch Facilities and Operations,* the authors describe one such overshadowed aspect: the building of the launch complex at Cape Kennedy, Florida. The book narrates how this important part of the lunar program, without which the *Apollo* missions could not have been launched, was developed and the impact it had on life in its neighborhood.

Bond, Peter R. **Heroes in Space: From Gagarin to** *Challenger.* New York: Blackwell, 1987. 467p. ISBN 0-631-15349-7.

"It was," says Peter Bond in the preface to his book, "my intention to write a history of manned spaceflight as a tribute to the courageous pioneers who volunteered to venture beyond Earth's atmosphere for the first time. The tragic events of 28 January 1986 simply reinforced my commitment to focus attention on the human aspects of space travel rather than the technology." This book presents the human side of manned space exploration, both Soviet and American, up to the *Challenger* disaster of 1986.

Brooks, Courtney G., James M. Grimwood, and Loyd S. Swenson, Jr. **Chariots for Apollo: A History of Manned Lunar Spacecraft.** Washington: U.S. GPO, 1979. 540p. Stock #3300-00768-0.

A nontechnical, illustrated history of the development of the *Apollo* spacecraft and the procedure of putting it into operation. The *Apollo* spacecraft had three sections—the command module, the service module, and the lunar module—and each section is described in this book.

Cassutt, Michael. **Who's Who in Space: The First 25 Years.** New York: G. K. Hall & Co., 1987. 311p. ISBN 0-8161-8801-7.

A collection of biographies of all those who have either traveled in space or were selected to travel until the *Challenger* disaster of January 28, 1986. The biographies are grouped into U.S., Soviet, and international astronauts. The U.S. section is further divided into NASA astronauts, civilian specialists, military payload specialists, and pilots of the X class of aircraft who were the precursors of manned space flight. The biographies concentrate on aerospace achievements, with little information on astronauts' personal lives. In essence, this book provides a history of the first 25 years of space flight as seen through the professional lives of those who flew the craft.

Clark, John D. **Ignition! An Informal History of Liquid Rocket Propellants.** Rutgers, New Jersey: Rutgers University Press, 1972. 214p. ISBN 0-8135-0725-1.

This book explains why particular fuels are used to reach space and how these fuels work to get us there. The book is written in a nontechnical manner that is easily understood.

Cooper, Henry S. F., Jr. **Before Lift-Off: The Making of a Space Shuttle Crew.** Baltimore, MD: Johns Hopkins University Press, 1987. 270p. ISBN 0-8018-3524-0.

This is the story of the crew of shuttle mission STS 41G, from the crew's selection through training to lift-off a year later, on October 5, 1984. To write this book, Cooper, who had already covered the U.S. space program for two decades, was allowed complete access to the crew by NASA. The book gives an interesting insight into the grueling training that astronauts have to endure before a space flight.

————. **Thirteen: The Flight that Failed.** New York: Dial, 1973. 199p. ISBN 0-8037-8765-0.

On April 13, 1970, the U.S. lunar mission *Apollo 13* was 72,000 kilometers (45,000 miles) from Earth when its oxygen tank number 2 exploded. On board were astronauts Fred Haise, James Lovell, and John Swigert. The astronauts' ingenuity and the ground crews' calm and persevering efforts to save their astronauts brought the three back alive to Earth. This book recounts step by step the dramatic flight of *Apollo 13*.

Cortright, Edgar M. **Apollo Expedition to the Moon.** Washington: U.S. GPO, 1976. 313p. Stock #3300-00630-6.

This book is an illustrated collection of chapters written by the people who created and flew the *Apollo* lunar missions. Dr. Wernher von Braun, for example, the brain behind the giant Saturn rocket motor that boosted the *Apollo*s into orbit, wrote the chapter on the Saturn rockets; Rocco Petrone wrote about the "Moonport" at Cape Kennedy; and astronauts described their flights. Excellent reading to discover what the Apollo project was truly all about.

Dunne, James A., and Eric Burgess. **The Voyage of *Mariner 10:* Mission to Venus and Mercury.** Washington: U.S. GPO, 1978. 224p. Stock #3300-00710-8.

On November 3, 1973, the United States launched *Mariner 10,* the first U.S. space probe bound for Venus and the first probe of humankind to use the slingshot effect—that is, using one planet's gravity to hurl an object toward another planet. *Mariner 10* flew by Venus and then used

Venus's gravity to travel on to Mercury. This book is an illustrated account of *Mariner 10*'s historic voyage and the wealth of information that *Mariner 10* radioed back to Earth.

Ezell, Edward Clinton, and Linda Newman Ezell. **The Partnership: A History of the Apollo-Soyuz Test Project.** Washington: U.S. GPO, 1978. 560p. Stock #3300-00730-2.

On July 17, 1975, the first international handshake occurred in space: *Apollo 18* of the United States and *Soyuz 19* of the U.S.S.R. docked together in space, and the two crews shook hands. The mission resulted from the 1972 summit on cooperation in space between U.S. President Richard Nixon and Soviet Prime Minister Alexei Kosygin. This book describes the project and how it was given shape and finally realized.

Fimmel, Richard O., Eric Burgess, and William Swindell. *Pioneer* **Odyssey: Encounter with a Giant.** Washington: U.S. GPO, 1975. 300p. Stock #3300-00662-4.

On March 2, 1972, the United States launched humankind's first interstellar space probe, *Pioneer 10*. It became the first space probe to leave the inner Solar System, after which it crossed the asteroid belt and headed for its first destination: Jupiter. It arrived at Jupiter on December 3, 1973, and radioed back to Earth humankind's first close-up look at the giant planet. This book recounts *Pioneer 10*'s mission from inception to its encounter with Jupiter.

Fimmel, Richard O., James Van Allen, and Eric Burgess. *Pioneer:* **First to Jupiter, Saturn and Beyond.** Washington: U.S. GPO, 1980. 300p. Stock #3300-00805-8.

The first interstellar spacecraft launched by humankind were the U.S. *Pioneers*, namely *Pioneer 10* and *Pioneer 11*. The space probes flew by Jupiter and Saturn and then plunged on into deeper space, heading out of the Solar System and into the Milky Way Galaxy. Bound for the stars, the probes carried messages from Earth for another civilization that may stumble upon these first goodwill ambassadors from Earth. The above book is a chronicle of the Pioneer project, starting from the purpose of the two missions and including descriptions of the space probes themselves, their designs and functions, the thought work that gave them shape, the data and pictures they sent back to Earth, and the people involved throughout the project.

Froehlich, Walter. *Apollo Soyuz.* Washington: U.S. GPO, 1977. 132p. Stock #3300-00652-7.

This is the story of the events that led to the conception of the first international space mission and the mission's fulfillment in space with the docking of *Apollo 18* of the United States and *Soyuz 19* of the U.S.S.R. The book discusses the implications of the mission and its impact upon the two countries and the world.

Furniss, Tim. **Manned Spaceflight Log.** New York: Jane's Publishing Co., 1986. 160p. ISBN 0-7106-0402-5.

As the name of the book suggests, this is a log of all manned space flights, of both the United States and U.S.S.R., up to the U.S.S.R.'s *Soyuz T15* mission of March 13, 1986. For each manned mission, the book contains data such as dates, crew names, spacecraft, duration of flight, purpose of flight, and whether or not the flight succeeded in accomplishing its mission. A very comprehensive chronicle of manned space flight.

Green, Constance McLaughlin, and Milton Lomask. **Vanguard: A History.** Washington: Smithsonian Institution, 1971. 309p. ISBN 0-87474-112-2.

The U.S. exploration of space started with the class of satellites called Vanguard. The above book is a history of the Vanguard program, its goals, its politics, its people, and its achievements.

Hacker, Barton C., and James M. Grimwood. **On the Shoulders of Titans: A History of Project Gemini.** Washington: U.S. GPO, 1977. 645p. Stock #3300-00643-8.

All maneuvers that were required to achieve a Moon landing were first tried and perfected by the ten two-man *Gemini* missions, launched between March 23, 1965, and November 11, 1966. This book is a history of the Gemini program, its frustrations, its accomplishments, the people who gave the missions shape from the ground, and the astronauts who flew the missions. The book also touches upon the building of the NASA space port at Houston, Texas, now called the Lyndon B. Johnson Space Center.

Harris, Leonard A. **Technology and the Civil Future in Space.** 26th Goddard Memorial Symposium, Volume 73, Science and Technology Series. San Diego: Univelt, 1989. 246p.

Based on the 26th Goddard Memorial Symposium held March 16–18, 1988, at NASA Goddard Space Flight Center, this volume includes sections on technology policy and plans, technology needs, and technology directions for the civil space program. Also included is a student program titled "Your Future in Space: Opportunities and Challenges," organized by panel members representing the civil, military, federal, industrial, and scientific communities.

Johnson, Nicholas L., and Darren S. McKnight. **Artificial Space Debris.** San Diego: Univelt, 1987. 110p. ISBN 0-89464-012-7.

The large quantities of artificial space debris resulting from fragments of rocket, spacecraft, and satellite break-ups that have accumulated in the past 25 years have become a serious hazard to future spacecraft. This book provides an overview of the problem of space debris. It defines and analyzes the space debris environment and discusses the hazards and issues of space debris.

Lay, Bierne, Jr. **Earthbound Astronauts: The Builders of** *Apollo-*Saturn. New York: Prentice-Hall, 1971. 198p. ISBN 0-13-222307-4.

The *Apollo* Moon missions could not have succeeded without the powerful Saturn V rocket motor that boosted the heavy *Apollo* payloads into orbit. This book is a history of the Saturn rocket—its conception, its development, and its testing—told through the stories of the people who engineered it.

Lewis, Richard S. **The Voyages of** *Apollo:* **The Exploration of the Moon.** New York: Harper & Row, 1974. 308p. ISBN 0-8129-0477-X.

The United States launched seven landing expeditions to the Moon, six of which landed and one of which had to return home due to serious problems. The author views these journeys as personal adventures, mass communications events, and evolutionary episodes in our movement out of our earthly cradle.

Morrison, David, and Jane Samz. **Voyage to Jupiter.** Washington: U.S. GPO, 1980. 199p. Stock #3300-00696-0.

On September 5, 1977, the United States launched humankind's third interstellar space probe, *Voyager 1*. Its mission was to fly past Jupiter and Saturn and then plunge on beyond the Solar System into the Milky Way Galaxy. *Voyager 1* arrived at Jupiter on March 5, 1979, and sent back the first close-up pictures of the planet's moons and its great red spot. The above book is a chronicle of the *Voyager* mission and the scientific and astronomical dividends gained from it, discussed against the background of the earlier *Pioneer* missions that also flew past Jupiter.

Murray, Bruce C., and Eric Burgess. **Flight to Mercury.** New York: Columbia University Press, 1977. 162p. ISBN 0-231-03996-4.

Mariner 10 of the United States gave humankind its first close look at Mercury when the space probe went into a solar orbit and flew by Mercury on March 29, 1974. *Mariner 10* flew by Mercury twice more before it ran out of attitude-control fuel and lost radio contact with Earth. The book above is about the *Mariner 10* mission, written by one who was directly involved with it. A personal account, the book describes the launch of *Mariner 10* on November 3, 1973, and the probe's use of Venus's gravity to propel itself into a solar orbit for the Mercury fly-bys. The mission's events in space are described along with corresponding occurrences on Earth.

O'Leary, Brian. **Mars 1999, Exclusive Preview of the U.S.-Soviet Manned Mission.** San Diego: Univelt, 1987. 160p. ISBN 0-8117-0982-5.

A former astronaut who has written a number of space books, O'Leary peers into the future with this illustrated book and investigates the possibility of a joint U.S.-U.S.S.R. manned venture to Mars. He envisions a fly-by of Venus, a rendezvous with Phobos (one of Mars' two moons) in 1998, and a landing on Mars in 1999.

Pellegrino, Charles R. **Chariots for Apollo: The Making of the Lunar Module.** New York: Atheneum, 1985. 238p. ISBN 0-689-11559-8.

The *Apollo* spacecraft, designed for manned lunar missions, had three sections: the command module, the service module, and the lunar module. The development of the section designed to land on the Moon, the lunar module, is recounted in this book. The book narrates how the lunar module was conceptualized, designed, developed, built, tested, and finally flown on its historic flights to the Moon. Also included are the views, frustrations, and accomplishments of the people involved with the module.

Schirra, Walter M., Jr., and Richard N. Billings. **Schirra's Space.** San Diego: Univelt, 1988. 238p. ISBN 1-55770-036-6.

Walter Schirra is the only astronaut who participated in the Mercury, Gemini, and Apollo programs. This autobiographical work tells the story of astronaut Schirra from his time in the navy through the Mercury and Gemini programs and then his participation in the Apollo program. He offers his assessment of the space program's past and present, his thoughts on the *Challenger* disaster, and his views of the current space flight challenge.

Sheffield, Charles, and Carol Rosin. **Space Careers.** San Diego: Univelt, 1984. 240p. ISBN 0-688-03256-7.

Until now there has been no book on space careers to answer the numerous queries received each year by the many societies involved in space exploration. This book covers the field so well that few questions remain unanswered. It provides not only a general background on space and the U.S. space program but also the specifics on training for a space career, how to get involved, and how to find a job. A full chapter is devoted to women in space, and space programs throughout the world are described. The book also has a directory of NASA facilities, space-oriented colleges, space societies and organizations, selected aerospace companies, state aviation and aerospace departments, international space organizations, space-related U.S. Senate and House committees with names of members, and much more factual information.

Simon, Michael. **Keeping the Dream Alive.** San Diego: Univelt, 1987. 224p. ISBN 0-915391-28-7.

This book deals with the space shuttle, the space station, and NASA's long-range objectives. It brings out the budget restrictions and lack of foresight that could cause future problems, as well as discussing space commercialization and how it can be financed. Successful space colonization, the author asserts, must have broad-based support including that of the government, and it must be the result of solid and practical ideas.

Stoiko, Michael. **Pioneers of Rocketry.** New York: Hawthorn, 1974. 129p. ISBN 0-8015-5875-X.

A book of concise biographies of the five men who were key to the development of rocketry and space flight. The men were William Congreve, England; Robert Esnault-Pelterie, France; Robert Goddard, United States; Hermann Oberth, Germany; and Konstantin Tsiolkovsky, U.S.S.R.

Wright, Pearce. **The Space Race.** New York: Gloucester Press, 1987. 32p. ISBN 0-531-17041-1.

A slim book that covers events from 1957 to the present, including the *Challenger* tragedy and its impact on U.S. space exploration. The book investigates how the space race began, which countries are currently involved, and where the race is likely to go. One section discusses the Strategic Defense Initiative, or "Star Wars."

Periodicals

This section lists only nontechnical periodicals dedicated to covering space issues that do not require membership in an organization to subscribe. Additional titles are listed within the "Directory of Organizations" (Chapter 5).

Air & Space
Smithsonian Institution
900 Jefferson Drive
Washington, DC 20560
Bimonthly. $18.

Founded in 1986, an excellent magazine for the young person and the layperson interested in space news and developments. The editorial office says about half the magazine is devoted to space-related issues, but this author finds an emphasis on air flight more than space flight. Subscription to the magazine automatically makes the subscriber an associate member of the National Air and Space Museum in Washington, D.C.

Space
Shephard Press, Inc.
Crystal Plaza 1, Suite 506
2001 Jefferson Davis Highway
Arlington, VA 22202
Bimonthly. $50.

A magazine, perhaps the only one, totally dedicated to space topics yet written in a language easily understood by both young people and laypeople. A recent issue contained articles such as "Materials for Space Station," "Space Reconnaissance—Soviet Style," and "Financing a Space Venture." The articles are vividly illustrated for clarity.

7

Films
and Videocassettes

THIS CHAPTER IS AN ANNOTATED list of films and videocassettes on space exploration. NASA is currently the foremost producer in the world, and the sole producer in the United States, of *original* film and videotape footage on space exploration. Films or videocassettes produced by other organizations are essentially derived from NASA's footage; therefore, this chapter lists only the original NASA releases. Annotations are courtesy of NASA or the National Audiovisual Center.

National Audiovisual Center Films and Videocassettes

The films and videocassettes listed in this section can be purchased from:

National Audiovisual Center
National Archives and Records Administration
Customer Services Section PY
8700 Edgeworth Drive
Capitol Heights, MD 20743-3701
(301) 763-1896

America in Space—The First Decade

Type:	16mm color film, 3/4" U-Matic, 1/2" VHS or Beta
Length:	28 min.
Date:	1963
Title Number:	130030/PY

Provides a ten-year history of NASA's role in the exploration of space and describes major accomplishments in aeronautics, atmospheric research, the use of scientific and applications satellites, studies of the Moon and planets, and manned space flight.

America in Space—The First Five Years

Type:	16mm color film, 3/4" U-Matic, 1/2" VHS or Beta
Length:	14 min.
Date:	1963
Title Number:	130040/PY

Traces the growth of the U.S. space program from *Explorer 1* through the initial phases of the Apollo program.

America's Journey to the Moon

Type:	16mm color film, 3/4" U-Matic, 1/2" VHS or Beta
Length:	29 min.
Date:	1974
Title Number:	007296/PY

Presents a documentary of man's first walk on the Moon.

Apollo 9—The Space Duet of *Spider* and *Gumdrop*

Type:	16mm color film, 3/4" U-Matic, 1/2" VHS or Beta
Length:	28 min.
Date:	1969
Title Number:	141371/PY

Presents a view of the *Apollo 9* astronauts James McDivitt, David Scott, and Russell Schweickart before, during, and after their Earth orbital mission. Concentrates on the launching, rendezvous, and docking of the command module—*Gumdrop*—and the lunar module—*Spider*—and the return and recovery of the crew.

Apollo 10—Green Light for a Lunar Landing

Type:	16mm color film, 3/4" U-Matic, 1/2" VHS or Beta
Length:	29 min.
Date:	1969
Title Number:	141373/PY

Shows highlights of the second lunar orbital mission by astronauts Tom Stafford, Eugene Cernan, and John Young in May of 1969. Features views of the Moon and Earth from space.

Apollo 12—**Pinpoint for Science**
Type:	16mm color film, 3/4″ U-Matic, 1/2″ VHS or Beta
Length:	28 min.
Date:	1969
Title Number:	141375/PY

Describes the second lunar mission with emphasis on scientific investigation and landing accuracy.

Apollo 13—**Houston, We've Got a Problem**
Type:	16mm color film, 3/4″ U-Matic, 1/2″ VHS or Beta
Length:	28 min.
Date:	1969
Title Number:	141375/PY

Describes the return flight of the crew of *Apollo 13* following an explosion on board the service module. Emphasizes the teamwork of mission control and the spacecraft crew as well as worldwide reaction to the crisis.

Apollo 14—**Mission to Fra Mauro**
Type:	16mm color film, 3/4″ U-Matic, 1/2″ VHS or Beta
Length:	29 min.
Date:	1971
Title Number:	141370/PY

Presents a documentary account of the mission including problems encountered on the way to the Moon and how they were solved. Depicts the activities during the scientific and geological traverses on the Moon and the return journey to Earth.

Apollo 15—**In the Mountains of the Moon**
Type:	16mm color film, 3/4″ U-Matic, 1/2″ VHS or Beta
Length:	28 min.
Date:	1971
Title Number:	141300/PY

Tells the story of the most ambitious and successful lunar landing mission to date. Includes details of the three lunar surface scientific expeditions, the experiments in lunar orbit, and the dramatic return to Earth.

Apollo 16—**Nothing So Hidden**
Type:	16mm color film, 3/4″ U-Matic, 1/2″ VHS or Beta
Length:	28 min.
Date:	1972
Title Number:	001989/PY

Presents a documentary account of the *Apollo 16* lunar landing mission and exploration in the Highland Region of the Moon, near the crater

Descartes. Through the use of cinéma vérité techniques, the real-time anxieties and lighter moments of the support teams were captured in mission control and the science support room.

Apollo 17—On the Shoulders of Giants
Type: 16mm color film, 3/4" U-Matic, 1/2" VHS or Beta
Length: 29 min.
Date: 1973
Title Number: 001992/PY

Documents the *Apollo 17* journey to Taurus-Littrow—the final lunar landing mission—and depicts the major events of the mission. Also describes the preparations for *Skylab*, the U.S.-U.S.S.R. link-up, and the space shuttle programs.

Apollo to the Moon
Type: 35mm color film
Length: 31 min.
Date: 1980
Title Number: A02482/PY

Describes the story of man's first landing on the Moon, from lift-off to recovery.

The Clouds of Venus
Type: 16mm color film, 3/4" U-Matic, 1/2" VHS or Beta
Length: 30 min.
Date: 1962
Title Number: 228050/PY

Documents the flight of *Mariner 2*, which was launched on August 27, 1962—our first effort to obtain scientific information about some of Venus's features that have been determined by astronomers. Presents the flight profile, trajectory, spacecraft features, and the experiments conducted throughout 109 days and 180 million miles of travel. A few preliminary scientific findings are summarized.

Conservation Laws in Zero-G
Type: 16mm color film, 3/4" U-Matic, 1/2" VHS or Beta
Length: 18 min.
Date: 1974
Title Number: 009614/PY

Presents Owen K. Garriott, *Skylab* on-board scientist, demonstrating angular momentum. Depicts astronauts spinning and rotating in space. Illustrates concepts through a spinning ice skater, a cat dropped with zero angular momentum landing on its feet, and a model of the *Explorer 1* satellite.

Eagle Has Landed—The Flight of *Apollo 11* (long version)
Type: 16mm color film, 3/4″ U-Matic, 1/2″ VHS or Beta
Length: 29 min.
Date: 1969
Title Number: 283380/PY

Shows humankind's first Moon landing, July 1969, including highlights of the flight of astronauts Neil Armstrong, Edwin Aldrin, and Michael Collins from launching through post-recovery activities, with emphasis on initial exploration of the lunar surface.

Fire to Space—The Story of *Centaur*
Type: 3/4″ U-Matic, 1/2″ VHS or Beta
Length: 29 min.
Date: 1982
Title Number: A08765/PY

Tells the story of the *Centaur* rocket, which boosted spacecraft and scientific probes to planets and moons and launched communications satellites. Describes how *Centaur* is carried to low-Earth orbit in the space shuttle cargo bay and then placed in orbit. *Centaur* is then remotely positioned and the rocket's engines are fired to propel a payload to its ultimate destination.

Great Spacecraft and Their Accomplishments
Type: 35mm color film
Length: 15 min.
Date: 1980
Title Number: A02483/PY

Discusses the accomplishments and scientific contributions of various spacecraft.

It Can't Happen to Me—The Anatomy of an Accident
Type: 16mm color film, 3/4″ U-Matic, 1/2″ VHS or Beta
Length: 23 min.
Date: 1987
Title Number: 002273/PY

Re-enacts an actual event at the Kennedy Space Center in Florida to illustrate how accidents can occur. Opens with a statement by the center director.

Jupiter Odyssey
Type: 16mm color film, 3/4″ U-Matic, 1/2″ VHS or Beta
Length: 28 min.
Date: 1974
Title Number: 009092/PY

Tells the story of the 620 million mile journey of *Pioneer 10* to Jupiter. Describes the 21-month trip during which *Pioneer 10* penetrated the unexplored asteroid belt without mishap, eliminating the long-held fear that high-speed particles or huge asteroids might destroy the spacecraft.

Magnetic Effects in Space

Type:	16mm color film, 3/4" U-Matic, 1/2" VHS or Beta
Length:	14 min.
Date:	1975
Title Number:	009954/PY

Illustrates basic principles of science by utilizing footage from *Skylab* in-flight science demonstrations. Astronaut Owen K. Garriott discusses several of the science experiments using film prepared from *Skylab* television transmission and demonstration equipment.

Magnetism in Space

Type:	16mm color film, 3/4" U-Matic, 1/2" VHS or Beta
Length:	19 min.
Date:	1975
Title Number:	009955/PY

Features astronaut Owen Garriott, who describes the basics of magnetism through demonstrations on Earth and in space on board *Skylab*. Reviews familiar magnetic effects and applications of magnets on Earth and shows how these effects are observable in space in new and different ways. Concludes with important present and future applications of magnetism in space.

A Man's Reach Should Exceed His Grasp

Type:	16mm color film, 3/4" U-Matic, 1/2" VHS or Beta
Length:	15 min.
Date:	1975
Title Number:	009488/PY

Presents the story of flight and of man's reach for a new freedom through aviation and the exploration of space. From the Wright Brothers' flight at Kitty Hawk to the landing on the Moon and future missions to the planets, this film depicts the fulfillment of the ancient dream of flight. Through the use of multiple images, the creative role of research is emphasized. Voices of scientists and statements by writers, poets, and philosophers document humankind's search for knowledge.

The Mission of *Apollo-Soyuz*

Type:	16mm color film, 3/4" U-Matic, 1/2" VHS or Beta
Length:	29 min.
Date:	1976
Title Number:	009970/PY

Stresses the spirit of cooperation and friendship that helped make the mission a success. Follows the mission time line and provides appropriate historical footage to detail the period of development and training. Concludes with a projection of the future of international cooperation in space, featuring the space shuttle and the European development of *Spacelab.*

Nineteen Minutes to Earth
Type: 16mm color film, 3/4" U-Matic, 1/2" VHS or Beta
Length: 15 min.
Date: 1977
Title Number: 010494/PY

Examines the scientific findings of the *Viking* mission to Mars. Viewers are introduced to a variety of information that was returned to Earth from Mars. Includes soil and atmospheric analyses and biological, meteorological, and geological data. Discusses difficulties encountered in interpreting *Viking* data. Actual photographs taken by the *Viking 1* and *Viking 2* spacecraft from orbit and on the surface of Mars are highlighted.

Orbiter Thermal Protection System
Type: 16mm color film, 3/4" U-Matic, 1/2" VHS or Beta
Length: 6 min.
Date: 1981
Title Number: A04572/PY

Describes how the thermal protection system shields the shuttle orbiter from high temperatures during re-entry and ascent, the types of insulation used on the orbiter, and the advantages of the silica fiber tiles.

Rocketry and Space Flight
Type: 35mm color film
Length: 21 min.
Date: 1980
Title Number: A02466/PY

Presents the history of rocket flight and introduces viewers to the pioneers of rocketry and space flight. Explores the nature of the rocket itself, revealing why it can travel in space, and describes how rockets maneuver in space.

Skylab
Type: 16mm color film, 3/4" U-Matic, 1/2" VHS or Beta
Length: 27 min.
Date: 1971
Title Number: 690109/PY

Shows the major objectives of the mission and its principal components. Also features the four launches involved and a few of the scientific investigations that will be performed.

Space Shuttle—A Remarkable Flying Machine
Type: 16mm color film, 3/4" U-Matic, 1/2" VHS or Beta
Length: 30 min.
Date: 1981
Title Number: A05752/PY

Describes in detail the first space shuttle flight. Shows the lift-off, in-flight activities, and landing at the Dryden Flight Research Center in California.

Space Shuttle—The Orbiter
Type: 16mm color film, 3/4" U-Matic, 1/2" VHS or Beta
Length: 14 min.
Date: 1981
Title Number: A04580/PY

Explains the space shuttle transportation system in general, including the main components: the orbiter, the external tank, and the solid rocket boosters. Also describes the size, configuration, and internal design of the orbiter and the myriad systems: propulsion, thermal protection, and computer.

Space Shuttle Communications
Type: 16mm color film, 3/4" U-Matic, 1/2" VHS or Beta
Length: 8 min.
Date: 1981
Title Number: A04568/PY

Explains the vast communications and tracking support for space shuttle operations, including the tracking and data relay satellite systems to be used for payloads, the NASA communications network, domestic satellites, and the deep space network for interplanetary missions.

Space Shuttle Extra Vehicular Activity
Type: 16mm color film, 3/4" U-Matic, 1/2" VHS or Beta
Length: 12 min.
Date: 1981
Title Number: A04582/PY

Explains the tools, tethers, space suits, life-support systems, and maneuvering units used during extra-vehicular tasks that are planned for shuttle operations. These tasks will help provide cost-effective, reliable, and sensible servicing operations for payloads.

Space Shuttle Propulsion
Type: 16mm color film, 3/4" U-Matic, 1/2" VHS or Beta
Length: 12 min.
Date: 1981
Title Number: A04588/PY

Explains the key propulsion systems, their use for shuttle operations, and how they differ from past expendable systems. These systems include the reusable solid rocket boosters, the external tank, the main propulsion system, the reaction control system, and the orbital maneuvering system.

The Time of Apollo
Type: 16mm color film, 3/4" U-Matic, 1/2" VHS or Beta
Length: 28 min.
Date: 1975
Title Number: 00961/PY

Presents U.S. President John F. Kennedy saying, in 1961, "This nation should commit itself to achieving the goal, before this decade is out, of landing a man on the Moon and returning him safely to Earth." Proves that Project Apollo has been successful.

NASA Films and Videocassettes

The following videocassettes can be rented or purchased from NASA. Videocassettes with production numbers starting with CL, JSC, VCL, and VJSC are available from:

NASA
Lyndon B. Johnson Space Center
Media Services Branch AP3
Houston, TX 77058

To purchase films with production numbers starting with HQ, write to the National Audiovisual Center at the address provided at the beginning of this chapter.

Videocassettes may also be rented from the NASA Regional Film Library serving your area. NASA's Regional Film Libraries and the areas they serve are listed below.

NASA
Goddard Flight Center
Public Affairs Office
Greenbelt, MD 20771
(301) 286-8101

Connecticut, Delaware, District of Columbia, Maine, Maryland, Massachusetts, New Hampshire, New Jersey, New York, Pennsylvania, Rhode Island, Vermont

NASA
Ames Research Center
Audiovisual Facility
918 N. Ringsdorf Avenue
Mountain View, CA 94043
(415) 694-6270

Alaska, Arizona, California, Hawaii, Idaho, Montana, Nevada, Oregon, Utah, Washington, Wyoming

NASA
Marshall Space Center
Public Affairs Office, CA-20
Marshall Space Center, AL 35812
(205) 544-6548

Alabama, Arkansas, Iowa, Louisiana, Mississippi, Missouri, Tennessee

NASA
Kennedy Space Center
Public Affairs Office
Code PA-EAB
Kennedy Space Center, FL 32899
(305) 867-7060

Florida, Georgia, Puerto Rico, Virgin Islands

NASA
Langley Research Center
Mail Stop 185—Technical Library
Hampton, VA 23655-5225
(804) 865-2634

West Virginia, Kentucky, North Carolina, South Carolina, Virginia

NASA
Lewis Research Center
Film Service Dept. 22
21000 Brookpark Road
Cleveland, OH 44135
(216) 433-5577

Illinois, Indiana, Michigan, Minnesota, Ohio, Wisconsin

NASA
Johnson Space Center
Public Information Branch AP3
Film/Video Distribution Library
Houston, TX 77058
(713) 486-9606

Colorado, Kansas, Nebraska, New Mexico, North Dakota, Oklahoma, South Dakota, Texas

Apollo 8: Go for TLI
Type:	16mm color film, 3/4" U-Matic, 1/2" VHS
Length:	22 min.
Date:	1969
Production Number:	JSC 500

Humankind's first voyage to another celestial body, including an orbit around the Moon on Christmas. Featured are air-to-ground tapes of astronauts' descriptions of the mission and outstanding photography of the Earth, the Moon, and on-board activity.

Apollo 11: For All Mankind
Type:	16mm color film, 3/4" U-Matic, 1/2" VHS
Length:	34 min.
Date:	1969
Production Number:	JSC 527

This film shows the first landing of humans on the Moon as the culmination of a dream. A new concept of reality begins as humankind stands on the threshold of a new age. Includes scenes showing launch, lunar orbit, landing, Moon walk, rendezvous, recovery, and return to Houston.

Assignment: Shoot the Moon
Type:	1/2" VHS, color
Length:	28 min.
Date:	1967
Production Number:	HQ 167

Summarizes the Moon exploration conducted by the unmanned *Ranger, Surveyor,* and lunar *Orbiter* space probes and shows how the data and photography contributed to the first manned flights to the Moon.

Birth of NASA Series

Episode 1: Moon; Episode 2: A Goal
Type:	3/4" U-Matic, 1/2" VHS, color
Length:	59 min.
Date:	1978
Production Number:	CMP 178

Episode 1: The beginning of NASA in 1958 and its early programs are highlighted. Because of a launch success rate of less than one-third, NASA starts a quality control program.

Episode 2: Although a fledgling organization, NASA reaches several milestones during 1960–1961. There are two highly successful unmanned orbital flights, the world's first weather and passive communications satellites, and two manned suborbital flights.

Episode 3: Around the World and on the Way; Episode 4: Preparing for the Moon

Type:	1/2″ VHS, color
Length:	59 min.
Date:	1986
Production Number:	CMP 179

Episode 3: John Glenn becomes an instant celebrity as the first U.S. citizen to orbit Earth.

Episode 4: Work continues on the liquid hydrogen-oxygen rocket. Tests of three Saturn rockets take place. Close-up Moon photographs taken by *Ranger 7* are studied, and plans are made for *Surveyor* to soft-land on the lunar surface.

Episode 5: Gemini: The Twins; Episode 6: Around the Moon

Type:	1/2″ VHS, color
Length:	59 min.
Date:	1986
Production Number:	CMP 180

Episode 5: During 1964–1965, two-man *Gemini* space flights provide scientists and astronauts with invaluable information and experience. The soft landing of *Surveyor 1* on the Moon in 1966 paves the way for a manned lunar landing.

Episode 6: In 1967, during a pre-flight test of the *Apollo* spacecraft, a fire erupts in the command module and kills three astronauts. As a result of the tragedy, the *Apollo* spacecraft is redesigned. In 1968, the Apollo program gains momentum with four flights—two unmanned and two manned. *Apollo 8* astronauts circle the Moon ten times.

Episode 7: Moon Landing; Episode 8: More Moon Exploration

Type:	1/2″ VHS, color
Length:	59 min.
Date:	1986
Production Number:	CMP 181

Episode 7: In 1969, humans first land on the Moon. With the beginning of the 1970s, NASA pursues its research in aeronautics and satellite technology.

Episode 8: The years 1972–1973 are busy for NASA. *Mariner 9* maps the entire surface of Mars, and *Pioneer 10* returns the first close-up pictures of Jupiter.

Episode 9: Transition Years; Episode 10: Shuttle Preparations and Planets
Type:	1/2″ VHS, color
Length:	59 min.
Date:	1986
Production Number:	CMP 182

Episode 9: The Apollo-Soyuz mission marks the first joint U.S.-U.S.S.R. space mission. Two *Viking* spacecraft land on Mars and conduct the first extensive search for life on that planet.

Episode 10: *Voyager 1* and *Voyager 2* are launched. Each space probe carries a copper record intended to serve as a greeting to other life forms.

Episode 11: Planetary Discoveries; Episode 12: Shuttle Era
Type:	1/2″ VHS, color
Length:	59 min.
Date:	1986
Production Number:	CMP 183

Episode 11: *Voyager 1* and *Voyager 2* fly by Jupiter, completing the first leg of their journey through the Solar System, and discover a thin ring around Jupiter and active volcanoes on Io. *Pioneer 11* sweeps by Saturn, giving us our first close look at this planet and its rings.

Episode 12: With the premier flight of the space shuttle *Columbia* in April 1981, the era of a reusable space transportation system begins. Seven months later *Columbia* is space-borne again. Three more shuttle flights follow in 1982.

Episode 13: Space Shuttle Matures
Type:	1/2″ VHS, color
Length:	30 min.
Date:	1986
Production Number:	CMP 184

In its 25th year—1983—NASA maintains its momentum of achievement. *Pioneer 10* becomes the first man-made object to leave the Solar System. *Challenger,* the second shuttle in a fleet to eventually number four, embarks on its first flight in April 1983.

Building towards New Heights
Type:	16mm color film, 3/4″ U-Matic, 1/2″ VHS
Length:	28.5 min.
Date:	1987
Production Number:	JSC 852

This production presents the work accomplished in orbit by the Skylab and space shuttle programs. Work shown includes testing of the Manned Maneuvering Unit on shuttle mission STS 41B, repairing the Solar Max satellite on mission STS 41C, retrieving two communication satellites on mission STS 51A, "fly-swatting" the satellite Syncom on mission STS 51D, and salvaging the Syncom on mission STS 51I.

Designing with Plaid
Type: 16mm color film, 3/4" U-Matic, 1/2" VHS
Length: 13 min.
Date: 1987
Production Number: VSC 1017R

Portrays the way NASA's Man-Systems Division uses computer graphics to analyze human maneuverability in the zero gravity of space. Using human-modelling software programs, artificial intelligence, and natural language interfaces, scientists can simulate astronauts working in space and design plans for a future space station.

Dream That Wouldn't Down
Type: 16mm black-and-white film, 3/4" U-Matic, 1/2" VHS
Length: 27 min.
Date: 1965
Production Number: HQ 125

The dream of Dr. Robert Goddard, the father of modern rocketry, is explored and examined through reminiscences of his wife, Mrs. Goddard. Included are historic scenes of Dr. Goddard's early experiments.

Four Rooms, Earth View
Type: 16mm color film, 3/4" U-Matic, 1/2" VHS
Length: 28 min.
Date: 1975
Production Number: HQ 239

Skylab was the first U.S. manned space station. This film is the story of the three missions, the nine astronauts, and their combined total of 171 days in *Skylab.*

Friendship 7
Type: 16mm color film, 3/4" U-Matic, 1/2" VHS
Length: 57 min.
Date: 1962
Production Number: JSC 124, HQ 059

John Glenn waited three years to orbit the Earth. Tracking stations are ready, and the world awaits the flight of *Friendship 7.* The spacecraft is launched and inserted into orbit. The first global astronaut from the

United States circles Earth every 88 minutes. His observations and reactions are shown while the rhythmic beat of his heart is heard in the background. Signs of impending disaster are relayed to John Glenn with recommendations for re-entry. *Friendship 7* encounters the searing heat of the atmosphere, which John Glenn describes as "a real fireball outside." The film ends with John Glenn's recovery from the Atlantic.

I Will See Such Things
Type: 3/4" U-Matic, 1/2" VHS, color
Length: 28.5 min.
Date: 1986
Production Number: CMP 186

In January 1986, *Voyager 2* flew past the planet Uranus and radioed thousands of images of the planet back to Earth. In this video, scientists at NASA's Jet Propulsion Laboratory analyze the images and discuss the knowledge gained from them.

It's You against the Problem
Type: 1/2" VHS, color
Length: 23 min.
Date: 1967
Production Number: HQ 146

Shows basic research in ablative materials being carried out by Dr. Simon Ostrach, director, Division of Fluid, Thermal, and Aerospace Sciences at Case Institute of Technology (now Case Western Reserve University), and by a graduate student working under Dr. Ostrach's guidance. Emphasizes the challenge of research and the education and life of a scientist.

Mars: The Search Begins
Type: 1/2" VHS, color
Length: 28.5 min.
Date: 1973
Production Number: HQ 236

The film depicts the planet Mars as known from the 7,000 pictures taken by the *Mariner 9* spacecraft. The Viking project storyline is carried by scientists Carl Sagan, Cornell University; Gerald Soffen, NASA; and Harold Masursky, U.S. Geological Survey.

Mercury: Exploration of a Planet
Type: 16mm color film, 3/4" U-Matic, 1/2" VHS
Length: 23.5 min.
Date: 1976
Production Number: HQ 282

The flight of the *Mariner 10* spacecraft to Venus and Mercury is detailed in animation and photography. Views of Mercury are featured. Included is animation on the origin of the Solar System. Jet Propulsion Laboratory Director Dr. Bruce C. Murray comments on the mission.

New View of Space

Type:	16mm color film, 3/4" U-Matic, 1/2" VHS
Length:	28 min.
Date:	1972
Production Number:	HQ 214

A dynamic overview of the space program—past, present, and future—that uses the underlying theme of photography to tell the story. A visual experience compiled from a collection of over nine million feet of film in the NASA film depository.

Nice Flying Machine

Type:	3/4" U-Matic, 1/2" VHS, color
Length:	8 min.
Date:	1984
Production Number:	CMP 124

Since the early days of the space program, astronauts have been trying to control their movement in the weightlessness of space. The Manned Maneuvering Unit (MMU) has now been developed and is used for extra-vehicular activity. In this tape, the development of the MMU is chronicled from prototype to current models. Projections are given for its use in the construction and maintenance of a manned space station.

Opportunities in Zero Gravity

Type:	16mm color film, 3/4" U-Matic, 1/2" VHS
Length:	18.5 min.
Date:	1976
Production Number:	JSC 684

Presents the characteristics of the weightless environment through the crews that manned *Skylab*. Early problems of working in weightlessness and the subsequent solutions are shown. Mobility, mass transfer, handling of small parts, and other aspects of zero gravity are demonstrated.

Partnership in Space: Mission Helios

Type:	1/2" VHS, color
Length:	27 min.
Date:	1975
Production Number:	HQ 254

This film follows the development and launch of *Helios,* a probe that orbits the Sun closer than any man-made object ever has. With a montage of art works depicting humankind's fascination with the Sun over the centuries, the film leads into present-day technological efforts to grasp the significance and influence of the Sun on our planet.

Portrait of Earth

Type:	16mm color film, 3/4" U-Matic, 1/2" VHS
Length:	27 min.
Date:	1981
Production Number:	HQ 299

This film explains satellites—what they are and how they perform their daily tasks in Earth orbit. In the fields of communication, meteorology, and Earth resources, satellites provide early warnings on hurricanes and forest fires and monitor pollution, marine resources, oceanography, land use, and agriculture.

Preparation and Packaging of Food for Space Flight

Type:	16mm color film, 3/4" U-Matic, 1/2" VHS
Length:	18 min.
Date:	1983
Production Number:	JSC 838

Depicts procedures documenting the preparation and packaging of food for shuttle missions.

***Skylab:* Space Station I**

Type:	16mm color film, 3/4" U-Matic, 1/2" VHS
Length:	28 min.
Date:	1974
Production Number:	JSC 651

Shows the launch of the unmanned *Skylab* space station on May 14, 1973, and the launches of the three manned missions: May 25, 1973, July 28, 1973, and November 16, 1973. Reviews the repair operation of the first mission, proving the presence of humans as vital to successful space exploration. Shows medical experiments looking at humans as they reacted under long-term weightlessness. Includes observations of solar flares, the Comet Kohoutek, and studies of the Sun for future energy. Narration is provided by each member of the three *Skylab* crews.

***Skylab:* The Second Manned Mission, a Scientific Harvest**

Type:	16mm color film, 3/4" U-Matic, 1/2" VHS
Length:	36.5 min.
Date:	1974
Production Number:	JSC 627

Shows the launch activities of astronauts Alan L. Bean, Owen K. Garriott, and Jack R. Lousma and their docking with the *Skylab* space station. Includes observations of student experiments, the Minchmog minnows and Arabella the spider, crew medical experiments and exercise routines, and the enabling of the Earth Resources Experiments Package. Shows manned operation of the Apollo Telescope Mount for observations of the Sun and beyond, space walks, testing of the Astronaut Maneuvering Unit, experiments to explore the industrial uses of space, and the *Skylab* living routine.

Space Navigation
Type: 1/2" VHS, color
Length: 21 min.
Date: 1967
Production Number: HQ 116

Explains in nontechnical terms the mathematical principles of charting a course in space for both manned and unmanned spacecraft. Shows navigational techniques to be considered in future flights from Earth to other planets, with emphasis on navigational problems and equipment used in the Apollo program.

Space Shuttle: A Remarkable Flying Machine
Type: 16mm color film, 3/4" U-Matic, 1/2" VHS
Length: 30 min.
Date: 1981
Production Number: JSC 814, HQ 318

Presents the hours prior to the first launch of the shuttle *Columbia*, its lift-off, the successful maiden voyage, and the landing at Rogers dry lake bed in California.

Space Station: The Next Logical Step
Type: 16mm color film, 3/4" U-Matic
Length: 15 min.
Date: 1984
Production Number: JSC 880

This production starts with a brief history of the U.S. space program and then emphasizes the space station as the next logical step in space exploration. It outlines the procedures for design and construction of the space station.

Space Station Assembly, Flights MB-1–MB-8
Type: 3/4" U-Matic, color
Length: 28 min.
Date: 1987
Production Number: VCL 1198

This series shows with computer animation and live shuttle scenes how the future space station will be assembled over eight shuttle flights. Each segment of the series (MB-1, MB-2, MB-3, etc.) covers one flight and demonstrates the construction of the respective space station element. The shuttle scenes include astronauts performing space walks and deploying a solar array and show the flight tele-robotic servicer attached to the shuttle arm.

Threshold of Opportunity, 1976

Type:	16mm color film, 3/4″ U-Matic, 1/2″ VHS
Length:	28 min.
Date:	1976
Production Number:	JSC 671R

Places the evolution of the space program into chronological perspective. The tension of the first launch is contrasted with the boldness of U.S. President John F. Kennedy's commitment to go to the Moon and return safely. After the lunar journey, humans turned earthward and began to concentrate on what could be learned in near-Earth orbit. The space shuttle is introduced as the culmination of the search for an economical and versatile space vehicle. The inception of an age of international cooperation is the concluding suggestion of the film.

Voyager

Type:	16mm color film, 3/4″ U-Matic, 1/2″ VHS
Length:	20 min.
Date:	1983
Production Number:	HQ 327

Depicts the exploration of Jupiter and Saturn by *Voyager 1* and *Voyager 2*. Photographs of the two planets and their moons are used with animation shots to show the varied surfaces of some of these moons. Highlighted are the major discoveries and a spacecraft problem during the Saturn flyby.

Glossary

ablation A process of heat dissipation by the continuous burning of material. During re-entry, a spacecraft produces tremendous heat due to friction with the atmosphere (*see* **aerodynamic heating**). The heat must be dissipated to protect the spacecraft. The dissipation is achieved by a heat shield, which consumes the heat by steadily burning up. By the time the entire shield is burned, the spacecraft has slowed and stopped generating heat. On the space shuttle, the heat shield is the black portion all around the bottom. The heat shield is composed of numerous tiles of ablative material, which have to be replaced for each flight.

abort The ending of a mission short of its objective. Usually an emergency procedure.

absolute altitude *See* **altitude**

acceleration The rate of change of velocity with respect to time. Acceleration can be linear or angular.

acceleration, axial Acceleration in the direction of a spacecraft's longitudinal axis. On the space shuttle, for example, axial acceleration is along the nose-tail axis.

acceleration, lateral Acceleration perpendicular to axial acceleration.

acceleration of gravity Acceleration caused by gravity. A body falling freely, such as a stone dropped from a height, is accelerating due to gravity. On Earth, at sea level, the acceleration of gravity is 981 centimeters (32.2 feet) per second every second of fall. Acceleration of gravity is generally denoted by "G."

accelerometer An instrument that measures the rate at which acceleration changes. The space shuttle has several accelerometers.

activated charcoal A charcoal made by removing hydrocarbons from organic substances. Often used in spacecraft cabins to absorb toxic gases and odors.

aerodynamic heating The heat generated by flying at high speeds through an atmosphere. The heat is generated by friction with the atmosphere. Aerodynamic heating is what happens when a spacecraft re-enters the atmosphere from space.

aeronautical mile *See* **nautical mile**

aeronomy The branch of science that studies the atmospheres of Earth and other celestial bodies.

aeropause The region of an atmosphere where the transition from the atmosphere to space occurs.

aerospace The entire realm of Earth's atmosphere and surrounding space.

aerospace medicine The branch of medicine that studies the effects of space flight on the human body.

afterbody A spacecraft sheds portions of itself when they are no longer needed for flight, and these shed portions may stay in orbit longer than the spacecraft. An afterbody is any portion that re-enters the atmosphere after the spacecraft.

afterburner A secondary combustion chamber that utilizes the heat exhausted by the primary chamber. Fuel injected into the secondary chamber burns from the exhausted heat. The afterburner is used for short bursts of additional thrust.

aft-firing thrusters Small rocket motors located at the tail of a spacecraft. Used for maneuvering.

agravic A region or state without gravity; the state of weightlessness.

air breakup The disintegration of a space vehicle by aerodynamic forces upon re-entry into the atmosphere. Air breakup is often deliberately planned into the vehicle's design so that the large parts of the vehicle break up into smaller parts and burn up upon re-entry instead of falling to the ground and possibly threatening life and property.

airlock A chamber that allows an astronaut to exit or enter a spacecraft in orbit without de-pressurizing the craft. The typical sequence of steps for exiting a spacecraft in orbit is: (a) astronaut, wearing a space suit, enters the airlock through the airlock's inner door; (b) the airlock is de-pressurized by the transfer of its air into the spacecraft; (c) the inner door is closed, which seals the spacecraft's atmosphere; (d) the airlock's outer door is opened into space, and the astronaut exits. The reverse sequence applies when the astronaut returns to the spacecraft.

algae Plants that readily photosynthesize. They use carbon dioxide and release oxygen, thus making them viable for air purification during long voyages on spacecraft. They are also a viable source for protein. Their use, however, is limited at present because they require the Sun's or similar light, and the equipment required to sustain them is bulky.

altitude The vertical distance of an object above the observer. The observer may be anywhere, on Earth or at any point in the atmosphere. Absolute altitude is the vertical distance to the object from an observer on the surface of Earth.

anergolic propellant A propellant in which the liquid fuel and liquid oxidizer do not burn spontaneously when they come in contact.

anhydrous Without water. For example, an anhydrous propellant works without using water.

aniline A liquid fuel compound with the chemical designation $C_6H_5NH_2$.

anti-G suit A garment for pilots to help them cope with the forces of high accelerations. The suit has inflatable bladders that, when inflated, prevent blood from pooling in the lower parts of the pilot's body, thus maintaining circulation in the pilot's upper body, including the brain. Without this circulation, the pilot would black out.

antigravity As yet a hypothetical field of energy that counters gravity.

aphelion For an object in an elliptical solar orbit, aphelion is the point on the orbit that is farthest from the Sun. Compare with **perihelion** and **apogee.**

apogee A term generally applied to artificial satellites. For a satellite orbiting in an elliptical orbit about a celestial body, apogee is the point on the orbit farthest from the body. Compare with **perigee** and **aphelion.**

apogee kick motor *See* **apogee motor**

apogee motor Also called apogee kick motor. It is a small rocket motor that fires at the apogee of a transfer orbit to insert a satellite into a geosynchronous orbit. First used on *Syncom 2* in 1963 to "kick" the satellite into a circular geosynchronous orbit, which has an altitude of 22,300 miles.

apolune A term applied to an elliptical lunar orbit. Apolune is the point on the orbit that is farthest from the Moon. *See also* **perilune.**

artificial gravity The centrifugal force produced by rotating a spacecraft. The force simulates Earth's gravity.

artificial satellite A man-made object in orbit around a celestial body.

astro- A prefix in English derived from the Greek *astron,* or star. U.S. space explorers are thus called astronauts.

atmospheric braking The slowing of a space probe by using a celestial body's atmosphere as a brake. The friction with the atmosphere slows down the probe. *See also* **braking ellipses.**

atomic rocket A nuclear rocket motor that uses atomic fission or fusion for propulsion. As yet this is only a proposed concept.

attitude The orientation of a spacecraft—its yaw, roll, and pitch—with respect to a particular frame of reference. In the case of spacecraft in Earth orbit, the frame of reference is usually Earth. However, interplanetary space probes often use stars as a frame of reference. *Mariner 4*, for example, used the star Canopus as a reference.

attitude control A system that changes or maintains a spacecraft's attitude. The system may comprise electronics as well as rocket motors.

attitude jets Fixed or movable gas nozzles or small rocket motors that adjust or change a spacecraft's attitude.

avionics The electronics and instrumentation that help in controlling a flight.

backout The reversing of a launch countdown.

biopak A pack carrying life-support equipment to sustain the life of a living being in space.

biosatellite A spacecraft designed to carry an animal life into space. Also, a satellite designed to conduct biological experiments in space.

bipropellant A rocket propellant made of two compounds that are separately fed into the combustion chamber.

blackout The loss of radio communications with a spacecraft. Blackouts may occur due to an emergency or due to natural phenomena. A natural breakdown, for example, occurs during re-entry into Earth's atmosphere. The re-entry creates atmospheric ionization that disrupts radio communications, causing a blackout. The term *blackout* also means loss of pilot consciousness due to high G-forces.

blast chamber *See* **combustion chamber**

blast deflector A structure that deflects the blast of a rocket motor. Deflecting the blast often becomes necessary to protect equipment and personnel.

blowdown The draining of a rocket motor's liquid propellants. Blowdown may be done before or after firing the motor.

blowoff The separation of a portion of a rocket or spacecraft by explosive force. The solid rocket boosters of the space shuttle, for example, separate from the shuttle by blowoff: Small explosives attached to the boosters explode, thus separating the boosters from the shuttle.

blowout disc A disc installed in the wall of a rocket's solid-propellant chamber to release excess pressure that may occur during the combustion of the propellant.

bonded grain Solid rocket propellant cast in a single piece and cemented or bonded to the motor casing.

boost The momentum given to a space vehicle during its flight, particularly during lift-off.

booster A secondary rocket motor, usually solid-fuel, that assists a main motor in propelling a spacecraft, particularly during lift-off. The space shuttle has two solid-propellant boosters that assist the sustainer engine, which is a liquid-propellant motor.

booster assembly The structure that supports one or multiple boosters.

braking ellipses A series of elliptical orbits that skim the atmosphere of a celestial body. Each time the space probe skims the atmosphere, friction with the atmosphere reduces the probe's orbital speed. Braking ellipses are used to slow down a probe for a landing. *See also* **atmospheric braking.**

braking rocket A rocket motor oriented to fire in a direction opposite to a spacecraft's motion. The opposite burn reduces the spacecraft's speed, thus acting as a brake.

bubble colony A proposed colony for the Moon or a planet within which will exist an Earth-like living environment, complete with humans, plants, and animals.

bungee An elastic cord that astronauts use in space to exercise and to secure equipment.

burning rate The linear measure of the amount of a solid propellant's grain that is consumed in a particular unit of time. Usually expressed in inches per second.

burnout The point at which a rocket motor's fuel exhausts and no more fuel combustion is possible.

burnout plug A valve in a rocket motor designed to retain a liquid fuel under pressure until the motor fires and provides the ignition flame.

burnout velocity A spacecraft's velocity at the time that burnout occurs.

burnout weight The weight of a spacecraft after burnout occurs. The weight includes any unusable fuel that may be left in the rocket motor's tanks.

capsule A spacecraft's cabin. Usually the only part of the spacecraft equipped for human, animal, or plant survival.

captive firing The firing of a rocket motor when it is restrained so that it cannot move. Usually a test firing.

capture The process in which a spacecraft comes under the influence of a celestial body's gravity.

carrier rocket A rocket vehicle that carries a payload, such as a satellite.

cast propellant A solid fuel fabricated by being poured into a mold and solidified.

celestial guidance A method of spacecraft guidance that uses stars as references for attitude and navigation.

chamber pressure The pressure of gases in the combustion chamber of a rocket motor, produced by fuel combustion. The motor's thrust is proportional to its chamber pressure.

characteristic length The ratio of the volume of the combustion chamber to the area of its nozzle's throat. Characteristic length provides a measure of the average distance that the burned fuel must travel to exit.

characteristic velocity An indicator of the amount of fuel a spacecraft would need for a flight to another celestial body. Characteristic velocity is the sum of all velocities that the spacecraft must attain throughout a voyage, and it is used to determine the amount of fuel the spacecraft must carry. For example, consider a spacecraft flying to the Moon. The velocities that the spacecraft must attain are:

escape from Earth	=	11.12 km (6.95 miles)/sec
Moon landing	=	3.00 km (1.88 miles)/sec
lift-off from Moon	=	3.00 km (1.88 miles)/sec
total (characteristic velocity)	=	17.12 km (10.71 miles)/sec

The spacecraft must carry enough fuel to achieve the above total, or characteristic, velocity if the spacecraft's rocket motor were to be fired only once.

chemical fuel Fuels that need an oxidizer for combustion. Nuclear fuels, in contrast, do not need an oxidizer. The space shuttle's fuels, both solid and liquid, are chemical fuels.

chest-to-back acceleration An acceleration in which the forces of acceleration are felt primarily on the chest, as would be felt in a sharply slowing automobile.

chlorella A unicellular algae considered to be suitable for attaining a closed environment that can sustain life in space. The algae uses photosynthesis to consume carbon dioxide and produce oxygen.

chuffing The irregular burning of a rocket fuel, applied mostly to solid fuels but also sometimes to liquid fuels.

circular velocity The velocity at which a spacecraft must move to maintain its orbit around a celestial body at a constant altitude, that is, a circular orbit.

circumlunar orbit A lunar mission in which the spacecraft's orbit encompasses the Moon.

circumplanetary space The space in the immediate vicinity of a planet. Circumplanetary space also includes the upper reaches of the planet's atmosphere.

circumterrestrial Something that encompasses the planet Earth, such as an orbit. The Moon's orbit is circumterrestrial.

circumterrestrial satellites Artificial—that is, man-made—satellites in Earth orbit. Such satellites may be manned or unmanned and include satellite debris.

cislunar satellite A satellite in Earth orbit that is operating at an altitude greater than about two Earth radii. *Cislunar* means between the Earth and the Moon. Compare with **terrestrial satellite.**

cislunar space The space between the Earth and the Moon.

close approach The time and place where the solar orbits of two planets are nearest one another.

closed ecological system A system for sustaining human and animal life in space by repeated recycling of carbon dioxide, urine, and other waste matter into oxygen and food. The recycling may be achieved through plants or chemical processors or both. Extended periods of manned travel in space will require such systems unless other means of sustaining life are invented.

cluster Two or more rocket motors forming a single propulsion system.

coasting flight A spacecraft's flight between the cutoff of one propulsion stage and the ignition of the next. Between the two stages, the spacecraft coasts along due to the momentum gained from the previous stage. Coasting is also the portion of flight between the end of the final burn and the reaching of the peak altitude.

cold-flow test A test of a liquid-fuel engine without igniting the fuel. The purpose of the test is to check fuel-flow systems, including tank pressurization.

colloidal propellant A solid propellant in which the mixture of fuel and oxidizer is so fine as to form a colloid. Also, a colloidal propellant may have the fuel and oxidizer atoms in the same molecule.

combustion, incomplete A state in which not all the fuel in the combustion chamber burns. Incomplete combustion may result from inadequate chamber design, or it may be deliberately designed into the system so that the unburned fuel acts as a chamber coolant. Generally, incomplete combustion denotes a system not functioning efficiently.

combustion chamber The chamber of a rocket motor, whether solid- or liquid-fuel, in which the fuel burns. A part of the combustion chamber is the nozzle, through which the exhaust gases exit and thereby provide propulsive thrust.

combustion efficiency The ratio of the energy actually released by the fuel during combustion to the energy contained in the fuel. A perfect combustion would release all the energy a fuel contains, in which case combustion efficiency would be 1, or 100 percent.

combustion limit In solid-fuel rocket motors, the combustion limit is the lowest pressure in the combustion chamber at which a given nozzle will support the burning of fuel without chuffing.

command destruct A self-destruct system in rockets that can be triggered remotely if the rocket goes astray. A rocket gone astray could threaten life and property on the ground, for which reasons it is then ordered to self-destruct by the range safety officer.

companion body A portion of a spacecraft or a payload, such as the last stage of a rocket or a discarded part, that orbits unattached to but along with its parent.

contamination The deposit of terrestrial microbes on another celestial body, such as the Moon. Disease germs transported to space colonies from Earth, for example, could threaten life. The process of introducing such microbes to hitherto clean celestial environments is called contamination.

contamination, back The contamination of Earth with possible microbial life from other celestial bodies. The alien microbes may not survive on Earth, but whether or not they survive, they could still threaten life on Earth.

contraorbital direction The direction opposite to the direction of travel of another body in the same orbit.

control system The electronics used to control a spacecraft's attitude and stability.

controlled leakage system A leakage that results from a deliberate design feature rather than from a flawed system. Carbon dioxide, for example, may be released from a spacecraft cabin through controlled leakage.

cooling, regenerative In liquid-fuel rocket motors, a method of using fuel to cool the motor's combustion chamber while at the same time pre-heating the fuel before combustion. In regenerative cooling, the fuel flows through coils coiled around the outside of the combustion chamber. While flowing through the coils, the fuel absorbs some of the chamber's heat, thus cooling the chamber and pre-heating itself. Pre-heating the fuel increases combustion efficiency and hence thrust.

coriolis effect The deviation in the flight path of a spacecraft caused by Earth's rotation. Over the Northern Hemisphere, the deviation is to the right; over the Southern Hemisphere, the deviation is to the left.

countdown The counting down of time before lift-off. Lift-off is usually time 0, or T-0:00:00. The countdown up to T-0:00:00 proceeds in reverse numerical order, such as T-0:00:10, T-0:00:09, T-0:00:08, and so on to T-0:00:00. All events leading to the lift-off occur at specific points along the countdown. T-0:00:00 is not necessarily the lift-off point, however, as is the popular belief. The space shuttle, for example, typically lifts off at T+0:00:03.

cryogenic fuel Liquid fuel that has to be kept at very low temperatures. At higher temperatures, cryogenic fuels do not remain liquid but convert to gases. Liquid hydrogen is a cryogenic fuel; its boiling point, or the point at which it becomes gas, is -263° C (-422° F). Liquid oxygen is also a cryogenic fuel. It is actually an oxidizer for liquid hydrogen. Liquid oxygen's boiling point is -183° C (-297° F).

cutoff Shutting off an engine or all engines. The term *cutoff* is applied to an intentional cutting of engines or to the natural shutting down of engines due to fuel depletion.

dawn rocket A dawn rocket is one that is launched into Earth orbit in an easterly direction so that the Earth's velocity of rotation augments the rocket's velocity. The Earth augments the rocket's velocity by 30 kilometers (18.5 miles) per second. The augmenting reduces the amount of fuel that the rocket must otherwise consume to achieve that extra velocity. Similarly, a dawn rocket is also often launched at dawn so that the Sun's gravity, which at dawn is pulling from the east, further augments the rocket's velocity. This advantage would be lost in the case of a dusk rocket.

deceleration The slowing down of a spacecraft. Also called negative acceleration.

deep space The space beyond the Solar System. Also called inter-galactic space.

de-orbit burn The firing of a spacecraft's rocket motor against the direction of motion to reduce the spacecraft's speed in orbit. A de-orbit burn thus has a braking effect on the spacecraft. The reduction of speed puts the spacecraft in a lower orbit. If the lower orbit passes through the Earth's atmosphere, the spacecraft re-enters the atmosphere.

descent path The path followed by a spacecraft during descent to Earth, particularly after re-entry into the atmosphere.

direct ascent A lunar or interplanetary flight usually proceeds as follows: (a) the spacecraft is first put into a parking orbit around Earth after lift-off; (b) the lunar or interplanetary flight then starts from the parking orbit. All manned U.S. lunar missions, for example, started from a parking orbit. Direct ascent, however, does not make use of a parking orbit. A spacecraft in direct ascent shoots for its destination with a single burn of its engines, without going into a parking orbit. *Surveyor 1*, the first U.S. space probe to make a soft landing on the Moon, was launched on May 30, 1966, into a direct ascent to the Moon. A single, prolonged burn of *Surveyor 1*'s engines accelerated the probe to escape velocity for escaping Earth's gravitational grip.

directional balance The orientation of a rocket with respect to its vertical axis.

directional control The system or the process of manipulating the system that controls the direction of a spacecraft's flight.

docking The mating of two spacecraft in space. Docking was first achieved on March 16, 1966, when *Gemini 8* of the United States docked with an Agena target vehicle, an unmanned vehicle designed especially for docking practice.

downlink A broadcast from an orbiting spacecraft to Earth. *See also* **uplink.**

downrange The area on Earth over which a spacecraft travels after launch and before entering orbit.

dusk rocket *See* **dawn rocket**

eccentricity In the universe, all orbits are elliptical. Eccentricity is a measure of the elongation of an ellipse. Eccentricity varies from 0 to 1. An ellipse with eccentricity 0 is a circle. An ellipse with eccentricity 0.866 is twice as long as it is wide.

electric engine A system of propulsion that uses electrically charged particles, or ions, to accelerate spacecraft. The energy of the particles is feeble, so the engine may not suffice for launch from Earth, but it may prove feasible for interplanetary voyages after conventional fuels have launched the spacecraft into Earth orbit.

elevon A control surface on the space shuttle that is used after the shuttle has re-entered the Earth's atmosphere. The elevon controls the shuttle's roll and pitch, thus acting as a combination of an aircraft aileron and elevator.

elliptic ascent   The profile of the ascent of a spacecraft into Earth orbit.

elliptic velocity   The velocity of a spacecraft in an elliptical orbit.

environmental space chamber   A chamber that simulates different environments of space, such as atmosphere, pressure, temperature, etc. Used for training astronauts.

equatorial orbit   An orbit on the same plane as Earth's Equator.

escape energy   The energy required per unit mass of a spacecraft for it to escape Earth's gravity.

escape velocity   The speed that a spacecraft must attain to escape a planet's gravity. The escape velocities for the planets of our Solar System are listed below in kilometers per second, with the miles-per-second figure given in parentheses:

Mercury 3.20 (2)   Mars 4.96 (3.1)   Uranus 20.80 (13)
Venus 10.08 (6.3)   Jupiter 59.20 (37)   Neptune 24.00 (15)
Earth 11.12 (6.95)   Saturn 35.20 (22)   Pluto Unknown

escape-velocity orbit   The orbit achieved around the Sun by a spacecraft escaping from Earth's gravity.

exobiology   The branch of science that investigates living organisms on other planets.

extra-vehicular activity   The technical name used by the National Aeronautics and Space Administration (NASA) for a space walk.

extra-vehicular mobility unit   NASA's technical name for a space suit.

fallaway section   A section, such as the lower stage, of a spacecraft that is cast away during flight, especially during ascent. The space shuttle, for example, has three fallaway sections: two solid rocket boosters and the external tank.

firing chamber   *See* **combustion chamber**

flame bucket   A deep cavity constructed beneath launch pads to receive the exhaust of a rocket.

flight path   The path followed by the center of gravity of a spacecraft during flight, also called the spacecraft's trajectory. The path is tracked with reference to any point on Earth, or it may be tracked with reference to a star.

flight profile   Flight path of a spacecraft as viewed from one side, making apparent the altitude at any point along the path.

free fall   The fall of an object under the sole influence of gravity. A stone dropped from a height, for example, free falls to Earth.

fuel cell   A source of water and electrical energy in space. A fuel cell mixes oxygen and hydrogen in a controlled manner to generate electricity and water.

fuel sloshing   In weightless conditions, liquid fuel in a partially empty tank may move away from tank outlets. This is called fuel sloshing. Sloshing prevents the re-starting of a rocket motor in space—an emergency situation, especially for manned missions.

geostationary orbit   *See* **geosynchronous orbit**

geosynchronous orbit   Also called geostationary orbit. It is an Earth orbit with an altitude of 35,900 kilometers (22,300 miles). A satellite in geosynchronous orbit has an orbital period equal to that of the Earth's rotation—23 hours and 56 minutes. Most communication satellites are in geosynchronous orbit.

G-force   The force of gravity, which is the product of the mass of a body times the acceleration of gravity. Gravitational acceleration on Earth at mean sea level is 981 centimeters (32.2 feet) per second for every second of descent. A body subjected to this acceleration is often said to be under a force of 1G, or more simply the body's own weight. Astronauts often experience forces that are several G's, or several times their weight. For example, the peak acceleration that the space shuttle exerts on its crew is 3G, but this is less than half the acceleration that *Apollo* astronauts experienced in the 1960s and 1970s.

gimbal   The rotation of a rocket motor's nozzle to control the direction of thrust and hence help steer a spacecraft. The nozzles of the space shuttle's solid rocket boosters, for example, can gimbal up to 6 degrees.

glide angle   The angle, or slope, of glide during descent. The glide angle for the space shuttle during the final leg of its descent from orbit is 22 degrees.

glide slope   *See* **glide angle**

gravisphere   The sphere of domination of the gravity of a celestial body with respect to another body. The gravity of a body dominates to the edge of its gravisphere, after which the gravity of another body becomes the more dominant gravity.

gravity well   A hypothetical "well" of gravity out of which a spacecraft must climb to escape into orbit.

grayout   Due to prolonged high acceleration, astronauts may suffer from reduced flow of blood to the brain, causing temporary impairment of vision. This condition is called grayout. *See also* **blackout.**

greater-than-escape-velocity orbit The solar orbit a spacecraft attains when traveling at a velocity greater than the escape velocity of a planet.

guidance, midcourse Guidance of a spacecraft while it is en route to its destination. Midcourse corrections were made, for example, on lunar missions to keep the spacecraft on course.

hand-held maneuvering unit A small jet held by astronauts to maneuver in space during space walks. The hand-held maneuvering unit was used in some flights prior to the space shuttle. The unit used nitrogen as fuel, which was fed to the unit through the astronaut's umbilical cord from the spacecraft.

heat barrier The speed above which friction with an atmosphere may damage a spacecraft. The heat barrier is of particular importance during re-entry. Special cooling systems have to cool the spacecraft when it goes beyond the heat barrier speed.

heat shield A shield of ablative material provided on critical portions of spacecraft to protect them from the heat of re-entry. The space shuttle, for example, travels about 26,400 kilometers (16,500 miles) per hour at re-entry. At that speed, tremendous heat is produced due to friction with the atmosphere. The heat can raise the shuttle's outside temperature as high as 1,650°C (3,000°F). Without a heat shield, therefore, the shuttle and its crew would be endangered. The shuttle's heat shield is called its thermal protection system and is in general the black-colored underbelly of the spacecraft. *See also* **ablation.**

Hohmann orbit A least-energy path of transfer between two elliptical orbits. Spacecraft transferring from one elliptical orbit to another along a Hohmann orbit consume the least amount of fuel. Also called tangential ellipse.

hybrid rocket motor A rocket motor that uses two or more different types of propulsion systems for thrust. The space shuttle, for example, uses a hybrid system of solid and liquid rocket motors.

hybrid-propellant motor A rocket motor that uses two or more different types of propellants for thrust.

hypergolic fuel A fuel that ignites spontaneously upon contact with an oxidizer. The spontaneous ignition eliminates the need for an ignition system.

hypersonic A body is said to be hypersonic if it can travel at speeds in excess of five times the speed of sound, or mach 5.

ideal burning A term applied to solid propellants. Ideal burning occurs when the propellant burns such that the thrust produced by the rocket motor and the pressure of combustion remain constant throughout the burn period.

ideal rocket A hypothetical rocket motor that has a velocity equal to the velocity of gases at the nozzle exit—that is, a rocket whose reaction is 100 percent equal to its action. Such a motor would be 100 percent efficient.

igniter A device that initiates combustion in a rocket motor.

impedance of space Space, as does an atmosphere, resists the propagation of electromagnetic waves. The resistance is called impedence of space and is about 377 ohms.

impulse, total The product of the average thrust of a rocket motor and the burn time.

impulse-weight ratio A ratio used to measure the efficiency of fuels. The impulse-weight ratio is the ratio of the total impulse to the lift-off weight of a spacecraft.

inclination The angle between the plane of the Earth's Equator and the plane of a spacecraft's orbit.

inertia The tendency of a body to remain in its current state of rest or motion until an external force changes that state. If in motion, the body also tends to maintain the direction of motion.

inertial guidance system A system for spacecraft to determine their position. The system is based on the principle of inertia. Modern aircraft also use the system for navigation.

inertial orbit All bodies that are not powered follow an elliptical orbit, as defined by Johannes Kepler in his laws of celestial motion. The first law states that celestial, or inertial, orbits are elliptical.

inertial system An instrument that measures a spacecraft's travel by measuring the spacecraft's accelerations relative to Earth. The system must be carried on board the craft.

inferior planet A planet closer to the Sun than Earth is. Mercury and Venus, for example, are inferior planets.

inhabited vehicle A spacecraft carrying humans.

injection pressure In a liquid-fuel rocket motor, the pressure at which the fuel is injected into the combustion chamber.

injector In liquid-fuel rocket motors, the device that injects fuel into the combustion chamber. The injector also mixes and atomizes the fuel for efficient combustion.

intergalactic space The space between galaxies.

interplanetary dust The particles of matter in interplanetary space.

interplanetary space   The space beyond the Moon and to the limits of the Solar System.

interstellar probe   A space probe that escapes the Solar System and travels into the space beyond. Humankind has so far launched four interstellar probes: *Pioneer 10, Pioneer 11, Voyager 1,* and *Voyager 2,* all U.S. probes. All are currently hurtling through deep space into the Milky Way Galaxy, headed for the stars.

interstellar space   Space starting from the edges of our Solar System and extending to the edges of our Milky Way Galaxy. The space beyond is intergalactic space.

ion rocket   A rocket motor that uses ionic propulsion. *See also* **electric engine.**

ionic propulsion   A method of propulsion that uses accelerated ions for thrust. The concept is as yet not capable of providing thrusts necessary to launch a spacecraft into space from Earth, but it may be effective after the spacecraft has reached space. *See also* **electric engine.**

Keplerian trajectory   All unpowered bodies in space follow elliptical orbits. These orbits are also called Keplerian trajectories.

Kepler's laws   Laws that govern celestial motion. The laws are: (1) planetary orbit is elliptical, with the Sun at one focus of the ellipse; (2) planets sweep out equal areas in equal times; and (3) the square of the time of a planet's orbit is equal to the cube of a planet's average distance from the Sun. The laws also apply to unpowered spacecraft in solar orbit.

knot   A unit of speed measuring 1.1516 statute miles per hour.

launch pad   The structure from which a spacecraft is launched. The structure may be on Earth or in space.

launch vehicle system   The system of rocket motors that launches a spacecraft into space. On the space shuttle, for example, the launch system includes the main engines, which are liquid-fuel motors, and the boosters, which are solid-fuel motors.

launch window   The period during which a spacecraft must be launched for it to reach its destination at a given time.

less-than-escape-velocity orbit   The orbit achieved by spacecraft not intending to escape from Earth, such as the space shuttle and Earth-orbiting satellites.

leveled thrust   A rocket motor programmed to deliver thrust at a constant rate.

life support   The sustaining of human or animal life in space.

lift-off   The rising of a spacecraft off its launch pad at the start of its ascent into space.

liquid hydrogen   A highly potent liquid fuel obtained by cooling hydrogen gas until it becomes liquid at -263°C (-422°F). The space shuttle uses liquid hydrogen for its main engines. When oxidized with liquid oxygen, liquid hydrogen delivers about 40 percent more thrust per pound than other liquid fuels, such as kerosene.

liquid oxygen   A highly potent oxidizer obtained by cooling oxygen gas to -183°C (-297°F). The space shuttle uses liquid oxygen to oxidize the liquid hydrogen fuel in its main engines.

lithium hydroxide   A compound used to remove carbon dioxide from spacecraft cabins.

long-playing rocket   An Earth-orbiting vehicle estimated to stay in orbit for a long period, usually several years.

lunar module   A two-man spacecraft designed to take astronauts from a lunar orbit down to a lunar landing, then back up to the lunar orbit. The lunar module was used on the manned Moon missions listed in Table 8. NASA's technical name for the module is lunar excursion module.

lunar satellite   A space probe destined to orbit the Moon.

mach   A unit of speed named after the Austrian physicist Ernst Mach. The unit uses the speed of sound as a reference. Mach 1, therefore, is the speed of sound, which in air at sea level is 1,218 kilometers (761 miles) per hour. A vehicle traveling at mach 2 is traveling at twice the speed of sound. A vehicle traveling below mach 1 is said to be subsonic, a vehicle traveling between mach 1 and mach 5 is called supersonic, and a vehicle traveling above mach 5 is called hypersonic.

main stage   In a multistage space vehicle, the main stage is the stage that develops the greatest amount of thrust. It is usually the lowest or first stage of the vehicle.

majority-rule circuit   Critical electronic circuits of a spacecraft are triplicated for insurance against failure. The computer then chooses the two most correct signals coming from the three circuits. Thus, if one circuit fails, its output will automatically be ignored. Redundant circuits operating in such a manner are called majority-rule circuits.

maneuver, capture   The transition from an open to a closed orbit is executed by a capture maneuver. *See also* **orbit, open** and **orbit, closed**.

maneuver, correction   A change of orbit during a space flight to a path closer to a pre-chosen orbit.

maneuver, escape A maneuver executed to escape from a celestial body, such as would be executed to escape from Earth for the Moon.

maneuver controller The control that an astronaut manipulates to execute any of the three principal spacecraft maneuvers: yaw, roll, and pitch.

microgravity A term that describes the apparent weightlessness and very small G-forces produced in orbit. *See also* **weightlessness.**

micrometeoroids Microscopic particles in space. They travel at high velocities, which makes them potential hazards to spacecraft. Spacecraft thus have to be shielded from them.

monopropellants Liquid rocket-motor fuels that have the fuel and oxidizer mixed into one propellant. Monopropellants require less equipment to handle in rocket motors but are unstable because they can ignite prematurely.

motor chamber In liquid-fuel rocket motors, the motor chamber is the combustion chamber (*see* **combustion chamber**). In solid-fuel motors, the motor chamber is the chamber that contains the solid fuel. In either case, combustion occurs in this chamber.

multipropellant A fuel comprising more than two ingredients coming together and burning in the combustion chamber.

nautical mile A unit of distance equal to 1.15 miles.

navigational planets Planets used as aids in celestial navigation. They are Venus, Mars, Jupiter, and Saturn.

negative G Force of acceleration in a direction opposite to that of a force of gravity. Gravity on Earth, for example, acts from head to feet. Negative G on Earth, therefore, would act from feet to head.

normal lapse rate The rate at which atmospheric temperature decreases with increasing altitude.

nozzle, underexpanded A type of rocket motor nozzle in which the gases of combustion are exhausted such that when they exit the nozzle, they are still at a pressure higher than the ambient pressure, with the result that they continue expanding beyond the nozzle.

nozzle area ratio The ratio of a nozzle's throat area to the exit area. The ideal nozzle area ratio allows the gases of combustion to exit the nozzle at the same pressure as the ambient pressure.

nozzle efficiency The efficiency with which a nozzle converts the potential energy of a burned fuel into kinetic energy for thrust.

nozzle exit angle The angle of divergence of a nozzle.

nozzle exit pressure The pressure at which gases exit from a nozzle.

nozzle throat The narrowest part of a nozzle. Usually the area nearest the combustion chamber of a rocket motor.

nuclear rocket A rocket motor or a rocket with a motor that derives its thrust from the products of nuclear fission or fusion. As yet only a concept, it may be feasible for interplanetary voyages.

on-board guidance computer A computer used for the guidance and maneuvering of a spacecraft in orbit. Spacecraft in Earth orbit travel at speeds in excess of 25,600 kilometers (16,000 miles) per hour. At that speed, astronauts have no time to perform the complex calculations required for guidance and maneuvering in space. Therefore, an on-board guidance computer is a very necessary part of a spacecraft.

orbit, closed An orbit that goes around a celestial body is called a closed orbit. Shuttle flights, for example, have closed orbits.

orbit, disturbed The orbit of a vehicle in a nonhomogeneous gravitational field.

orbit, open An orbit that does not go around a celestial body is an open orbit. The *Voyager* space probes, for example, have open orbits.

orbital curve The trace of an orbit on a flattened map of the Earth or another celestial body about which the spacecraft is orbiting. Each successive orbit is displaced by the amount of rotation of the body between each orbit.

orbital element Any of six characteristics necessary to completely define a planetary orbit. The six elements are: (a) eccentricity, which defines the shape of the orbit; (b) semi-major axis, which fixes the size of the orbit; (c) inclination; (d) longitude of ascending node, which determines the direction in which the planet crosses the ecliptic from south to north; (e) longitude of perihelion, which determines how far around the orbit the point nearest to the Sun lies as measured from the ascending node; and (f) time of perihelion passage, which provides a starting point for reckoning, after which the speed and position of a planet can be calculated from Kepler's third law of planetary motion (*see* **Kepler's laws**).

orbital energy The sum of the potential and kinetic energies of a body orbiting another body.

orbital period The time taken by a spacecraft to complete one orbit.

orbital velocity The average velocity of a spacecraft in orbit.

orbiter The portion of a spacecraft that orbits a celestial body. In the case of the space shuttle, only the winged structure is the orbiter.

oxygen, regenerative A spacecraft's oxygen supply that is recycled for repeated use after being cleaned of carbon dioxide. This is accomplished by man-made chemical devices or by plant life on board. The latter is suggested for interplanetary manned missions, which will be of long duration.

paddle-wheel satellite A satellite with large solar panels deployed to generate electrical energy. The panels resemble giant paddles alongside the satellite, hence the name paddle-wheel satellite. *Explorer 6* of the United States, launched on August 7, 1959, was the world's first paddle-wheel satellite.

parking orbit An interim Earth orbit for spacecraft going beyond Earth. All manned lunar missions, for example, first went into a parking orbit for preparations and then went on to the Moon.

passive communications satellite A satellite that reflects a signal received from Earth but does not electronically process and re-transmit the message.

passive satellite A satellite that receives but does not transmit any messages. A satellite may be designed to be passive only over enemy territory, or it may be passive due to faulty circuitry on board.

PAW ascent PAW stands for "powered all the way." PAW ascent, therefore, is the ascent of a spacecraft that is powered all the way. PAW ascents follow the shortest distance—a straight line—between Earth and a point in Earth orbit. PAW ascents consume the most fuel. They are not considered feasible for orbits higher than 125 nautical miles altitude.

payload The freight, such as a satellite or spacecraft, carried by a space vehicle.

perigee A term generally applied to artificial satellites. For a satellite orbiting in an elliptical orbit about a celestial body, perigee is the point on the orbit that is nearest to the body. Compare with **apogee** and **perihelion.**

perihelion For an object in an elliptical solar orbit, perihelion is the point on the orbit that is nearest to the Sun. Compare with **aphelion** and **perigee.**

perilune A term applied to an elliptical lunar orbit. Perilune is the point on the orbit nearest to the Moon.

period, orbital The time taken by a spacecraft to complete one orbit around a celestial body.

perturbation The effect of the gravitational pull of one body upon the orbit of another. Perturbation causes departures from smooth orbits.

Earth, for example, weaves slightly along its elliptical path due to the perturbation caused by the Moon's gravity.

pitch A spacecraft maneuver about an axis transverse to the craft's longitudinal axis. On the space shuttle, pitch would be the dipping or raising of the shuttle's nose about an axis across its wings, which is perpendicular to its longitudinal—or nose-tail—axis.

pitch attitude The orientation of a spacecraft with respect to its pitch axis. *See also* **pitch.**

pitching moment The rising and falling of a spacecraft's nose. When the nose rises, the pitching moment is positive; when the nose drops, the pitching moment is negative, also called diving moment.

polar orbit An orbit that passes over or near Earth's poles.

positive G The opposite of negative G (*see* **negative G**). Earth's normal influence on an upright human is a positive G, with gravity acting down from head to feet.

powered landing The landing of a spacecraft onto a celestial body in which its rocket motors are fired as brakes to make a soft landing. All manned lunar missions, for example, were powered landings.

pressure feed system The device in a liquid-fuel rocket motor that forces fuel into the combustion chamber under pressure.

pressure limit The upper and lower limits of pressure within which a solid-fuel rocket motor delivers optimum performance.

primary The celestial body around which another body orbits. Earth, for example, is the Moon's primary. The Sun, likewise, is Earth's primary.

progressive burning In solid-fuel rocket motors, the burning of the fuel such that the chamber pressure steadily rises throughout the burn time, thus delivering steadily increasing thrust.

propellant A rocket motor's fuel. A propellant may be solid or liquid. The space shuttle, for example, uses both.

propellant erosion The damage done to a propellant by heat or other causes, leading to irregular burning of the propellant.

propulsion system The system that propels a space vehicle, comprising rocket motors (both solid- and liquid-fuel), liquid-fuel tanks, pumps, ignition devices, combustion chambers, etc.

radiator, space A radiator designed to dissipate heat in space.

range safety officer A person responsible for the safe launching of a spacecraft. The range safety officer also has the authority to order the remote destruction of an unmanned rocket if the rocket shows signs of flying out of control and threatening life and property on the ground.

recovery The process of recovering the spent part of a space vehicle after it falls back to Earth. The solid rocket boosters on the space shuttle, for example, fall back into the Atlantic, from where they are recovered and reused.

re-entry The entering of a spacecraft into Earth's atmosphere from space. This is a critical phase in a space flight, because tremendous heat is generated during re-entry due to friction with the atmosphere, and the heat causes a blackout of communications between Earth and the spacecraft.

re-entry body The portion of a spacecraft that re-enters the atmosphere at the end of a space flight. In the case of the *Apollo* spacecraft, it was the command module; in the case of the space shuttle, it is the orbiter.

regressive burning In solid-fuel rocket motors, the burning of the fuel such that the chamber pressure steadily decreases throughout the burn time, thus delivering steadily decreasing thrust.

remaining body The portion of a spacecraft that reaches space after all its launch stages have been used. In the case of the space shuttle, the remaining body is the orbiter.

remaining mass The mass of a spacecraft that reaches space after all its launch stages have been used and discarded.

resonant burning Unstable combustion of fuel in a solid-fuel rocket motor. The unstable combustion causes a screaming sound.

retro-fire *See* **de-orbit burn**

retrograde orbit An Earth orbit in a direction opposite to that of Earth's rotation. Earth's rotation is west to east; a retrograde orbit is, therefore, from east to west.

retro-rocket A rocket that produces thrust in the direction opposite to the motion of a spacecraft. A retro-rocket reduces a spacecraft's speed to bring it down to a lower orbit. If the lower orbit crosses Earth's atmosphere, the spacecraft re-enters the atmosphere. *See also* **de-orbit burn.**

roll The motion of a spacecraft about its longitudinal—or nose-tail—axis.

rotation The motion of a spacecraft that is a combination of all three basic spacecraft motions: yaw, roll, and pitch.

rudder A control surface used by the space shuttle to control yaw after it re-enters the atmosphere.

rumble Unstable combustion of fuel in a liquid-fuel rocket motor. The unstable combustion produces a low-pitched, low-frequency noise like a rumble.

satellite period The sidereal period of time taken by a satellite to complete one orbit. A sidereal period is one that is measured with respect to the stars.

sealed cabin The part of a spacecraft that supports life. It is sealed to preserve the life-supporting environment within. Also called the space cabin.

secondary propulsion system The propulsion system used by space-craft to maneuver while in orbit. The secondary propulsion system enables the spacecraft to move in six basic directions: up, down, left, right, forward, and reverse.

separation rocket Used in a multistage rocket motor, the separation rocket is a small, usually solid-fuel motor that helps separate a spent stage from the rest of the spacecraft. The separation rocket provides the space-craft extra thrust to shed a spent stage and speed away from it.

service tower The vertical structure that provides access to various levels of a spacecraft to prepare it for launch. The service tower also supports the spacecraft in a vertical orientation until the spacecraft's rocket motors and guidance control systems take charge.

shirt-sleeve environment An environment, such as in the space shuttle's cabin, that does not require space suits or other protective clothing for human survival.

shutdown The deliberate cutoff of an engine or cluster of engines in a stage.

solar cell Also called a photovoltaic cell, the solar cell is a source of electrical energy on Earth and in space. It converts sunlight to electricity.

solar sail Light exerts a very feeble force on whatever it falls on, and sunlight is no exception. The solar sail is a concept to harness the feeble force of sunlight into a propulsive force for spacecraft, much as a sailboat uses wind for propulsion.

space cabin *See* **sealed cabin**

space equivalence The upper reaches of Earth's atmosphere where conditions for survival are almost identical to those required for survival in space. Above about 915 meters (30,000 feet) altitude is generally considered the space equivalent for humans, for at such high altitudes humans need oxygen and special pressure-retaining clothing to survive—two conditions, for example, that are similar to the requirements for survival in space.

space station A proposed habitat in space to be used for various purposes such as scientific research, astronomy, and medicine and also for launching interplanetary space probes. The National Aeronautics and Space Administration (NASA) plans to put a space station into orbit in the mid-1990s. The space station will be placed in a low Earth orbit with average altitude of 500 kilometers (315 miles) and an inclination of 28.5 degrees to Earth's equatorial plane. The station will support a crew of eight. The crew will be changed every three months.

space suit A suit designed to support life in space. The suit's inner environment duplicates Earth's environment in pressure, temperature, humidity, etc. The suit also protects astronauts from micrometeoroids. The suit is worn primarily during space walks.

specific impulse A measure of the efficiency of a rocket motor, specific impulse is rocketry's equivalent of the automobile's miles per gallon efficiency. Specific impulse is expressed in seconds and it is the ratio of the rocket motor's thrust to the weight of the fuel burned in one second of burn time.

speed brake On the space shuttle, this device increases drag to slow down the shuttle after touchdown.

spin stabilization A method of stabilizing a spacecraft during its ascent from Earth by slowly rotating the craft about its longitudinal axis. The method is not used on modern spacecraft such as the space shuttle.

staging The separation of the spent stage of a rocket motor from the rest of the spacecraft. The spent stage is dropped and either falls back to Earth (during ascent) or drifts away in space.

stationary orbit *See* **geosynchronous orbit**

stay time A term applied to liquid-fuel rocket motors. Stay time is the average time spent by a gas molecule in a motor's combustion chamber before the molecule exits from the nozzle and produces thrust.

steering rocket A rocket motor or a set of motors provided on spacecraft to steer them in space. The motors may be liquid- or solid-fuel. The space shuttle's steering system comprises an orbital maneuvering system of rocket motors and a reaction control system made up of 44 small motors that are mounted on the nose and tail. The orbital maneuvering system allows the shuttle to make large orbital changes and also to de-orbit, and the reaction control system is used for maneuvers such as yaw, roll, and pitch. Both the orbital maneuvering system and the reaction control system use liquid fuel and oxidizer: The fuel is monomethyl hydrazine, and the oxidizer is nitrogen tetroxide.

step principle A design feature on some rockets in which one stage is mounted directly onto another. When a lower stage is used up, it is ejected, and the next upper stage takes over. With each stage's ejection, the weight of the spacecraft decreases, so the next stage has less work to do. The step principle was used on all *Apollo* spacecraft, for example.

step rocket A multistage rocket.

subgravity A gravitational force less than Earth's gravity, which is 1G.

suborbital flight A flight that reaches space but achieves only a partial orbit, not a full orbit. The first U.S. manned flight into space was a suborbital flight. It occurred on May 5, 1961, with astronaut Alan Shepard aboard.

subsatellite A portion of a satellite that has a mission objective of its own. Once in space, the subsatellite is ejected and assumes its own orbit.

Sun-synchronous orbit A polar, Earth orbit. In this orbit, the spacecraft remains in a constant time zone as it travels around the Earth.

superior planet A planet farther away from the Sun than the Earth is. The superior planets are Mars, Jupiter, Saturn, Uranus, Neptune, and Pluto.

surface gravity The rate at which a freely falling body is accelerated by gravity.

sustainer engine The rocket motor that stays with a spacecraft during ascent after the booster has dropped off. The sustainer motor sustains or steadily increases the spacecraft's velocity during ascent. On the space shuttle, the sustainer engine is the main engine, which sustains the shuttle's ascent after the solid rocket boosters are spent and have been dropped off.

synchronous satellite A satellite in geosynchronous orbit (*see* **geosynchronous orbit**).

takeoff mass The mass of a spacecraft at launch. The mass includes all stages, fuel, payload, etc.

takeoff weight The weight of a spacecraft at launch. The weight includes all stages, fuel, payload, etc.

tangential ellipse *See* **Hohmann orbit**

telemetry The data received electronically from a spacecraft during flight. Telemetry informs ground control about the condition of the crew and of various critical parts and functions of the spacecraft.

temperature limits The temperature extremes, upper and lower, within which a solid fuel burns stably. Above and below these extremes, the fuel's combustion is not reliable.

terrestrial satellite A satellite in Earth orbit, within an altitude of about two Earth radii.

throat The most constricted portion of a rocket motor's nozzle.

throatable A nozzle whose size and profile can be varied. A throatable nozzle can be especially useful in a solid-fuel rocket motor to maintain uniform thrust throughout the burn time of the fuel.

thrust The propulsive force produced by a rocket motor. The space shuttle has three main motors, each producing a rated thrust of 1.6 million newtons (375,000 pounds) at sea level.

thrust chamber *See* **combustion chamber**

thrust equalizer A safety device that prevents motion of a spacecraft if its solid-fuel rocket motor ignites accidentally. The device is usually a vent at the top of the combustion chamber that is left open until launch time. If the fuel ignites accidentally before launch, the gases of combustion will blow out from both the top and the bottom of the motor, thus equalizing thrust on both sides and preventing the spacecraft from launching prematurely.

thrust meter An instrument that measures a rocket motor's thrust.

thrust misalignment Thrust directed accidentally in an undesired direction. Thrust misalignment can have serious consequences, especially during the initial stages of a spacecraft's ascent into orbit.

total impulse The total thrust produced by a rocket motor during its entire burn time.

tracking The process of keeping a track of the position of a spacecraft at all times. Tracking can be accomplished by Earth-based radar systems or by homing in on signals transmitted by the spacecraft.

transfer ellipse The path traced by a spacecraft as it changes orbits from one elliptical orbit to another. *See also* **Hohmann orbit.**

transfer orbit *See* **transfer ellipse**

translation The motion of a spacecraft along its principal axis. For the space shuttle, for example, translation would be motion along its nose-tail axis.

transverse acceleration Acceleration acting transversely to the longitudinal or principal axis of a spacecraft. On the space shuttle, for example, transverse acceleration would be at right angles to the nose-tail axis. On human bodies, transverse acceleration would be from chest to back or vice versa instead of from head to toe.

trapped propellant In a liquid-fuel rocket motor, the amount of fuel left in the tanks that cannot be used because of the suction limitations of the pumping systems.

24-hour orbit *See* **geosynchronous orbit**

ullage In liquid-fuel rocket motors, the tanks are not filled to the brim to allow for the fuel's expansion due to heat. The amount of fuel that can be accommodated in the space left vacant is called ullage.

umbilical cord The cord that attaches a space-walking astronaut to his spacecraft. The cord conveys oxygen and communications to the astronaut, and it also carries nitrogen fuel for the astronaut's hand-held maneuvering units. Modern space suits, however, do not need an umbilical cord, for they can independently support the astronaut's life and communications.

unrestricted burn In solid-fuel rocket motors, unrestricted burn is achieved by boring a hole in the fuel along the motor's longitudinal axis, so that the entire length of the fuel chamber burns simultaneously. This method of burning produces a great amount of thrust for a short time.

uplink A broadcast from Earth to an orbiting spacecraft. *See also* **downlink.**

vector steering A method of steering in which combustion chambers of rocket motors are mounted on gimbals so that the thrust can be directed in order to steer the spacecraft in a desired direction. *See also* **gimbal.**

vernier engine A very small rocket motor used for fine adjustments or alignments of a spacecraft's yaw, roll, and pitch. The space shuttle, for example, has six vernier engines, each producing 110 newtons (25 pounds) of thrust.

vernier engine cutoff The shutting down of a vernier engine.

weight flow rate In liquid-fuel rocket motors, weight flow rate is the rate, expressed in weight per unit of time, at which the fuel flows into the combustion chamber.

weightlessness The condition of zero gravity or microgravity, produced by balancing a gravity with the centrifugal force generated by an orbiting spacecraft.

wet-fuel rocket Another name for a liquid-fuel rocket motor.

white room The room in which astronauts prepare for a space flight before entering the spacecraft. The name "white room" is borrowed from a similar term used for clean rooms—free of dust and other contamination—in industries and hospitals.

yaw The left or right motion of the nose of a spacecraft.

zero gravity   *See* **microgravity** or **weightlessness**

zero stage   The common name given to a cluster of solid-fuel rocket motors that provide additional boost—in addition to a sustainer engine's thrust—to a spacecraft during ascent into orbit. The zero stage helps the spacecraft carry a greater payload.

Index

Heterick Memorial Library
Ohio Northern University

DUE	RETURNED	DUE	RETURNED
1.		13.	
2.		14.	
3.		15.	
4.		16.	
5.		17.	
6.		18.	
7.		19.	
8.		20.	
9.		21.	
10.		22.	
11.		23.	
12.		24.	